Computational Fluid Dynamics

Stefan Lecheler

Computational Fluid Dynamics

Getting Started Quickly With ANSYS CFX 18
Through Simple Examples

 Springer

Stefan Lecheler
Faculty of Mechanical Engineering
Bundeswehr University Munich
Neubiberg, Germany

ISBN 978-3-658-38452-4   ISBN 978-3-658-38453-1   (eBook)
https://doi.org/10.1007/978-3-658-38453-1

This book is a translation of the original German edition "Numerische Strömungsberechnung" by Lecheler, Stefan, published by Springer Fachmedien Wiesbaden GmbH in 2018. The translation was done with the help of artificial intelligence (machine translation by the service DeepL.com). A subsequent human revision was done primarily in terms of content, so that the book will read stylistically differently from a conventional translation. Springer Nature works continuously to further the development of tools for the production of books and on the related technologies to support the authors.

This Springer imprint is published by the registered company Springer Fachmedien Wiesbaden GmbH, part of Springer Nature.
The registered company address is: Abraham-Lincoln-Str. 46, 65189 Wiesbaden, Germany

Preface

In the meantime, commercial flow calculation programs such as CFX, FLUENT, and STAR-CD have reached such a high level of development in terms of flexibility, accuracy, and efficiency that they are used for very many flow problems in industrial applications. Without them, efficient development of new vehicles, aircraft, engines, and turbines would no longer be possible.

For this reason, development engineers must nowadays be proficient in the use of CFD (computational fluid dynamics) programs. Since even small input errors can have a major impact on the calculation result, these must be avoided at all costs. This book aims to contribute to this and, in addition to the most important theoretical principles, above all to provide numerous tips and experiences relevant to practice.

This book originates from the lecture "Numerical Flow Calculation," which is offered in the master's program in computer aided engineering at Bundeswehr University Munich. Within this program students learn by means of practical exercises with a commercial flow calculation program how to use it, which theory is behind it and how to interpret the results. Their suggestions were incorporated into the book. For example, complicated mathematics was avoided as far as possible and great importance was attached to clarity and comprehensibility.

I would like to thank all supervisors and colleagues who supported me in my work in the field of computational fluid dynamics at the Institute of Space Systems at the University of Stuttgart, at the von Karman Institute in Brussels, in the Thermal Machines Laboratory at ABB Turbo Systems in Baden/Switzerland, in gas turbine development at ALSTOM Baden/Switzerland, and at Bundeswehr University Munich.

I am very pleased that the fourth edition can now be published. In it, a few corrections have been made, and the three exercise examples have been adapted to the latest version ANSYS18.1. The CAD files are now read directly as in real applications. These are available on the Internet at www.unibw.de/mb/institute/we5/we51/downloads/downloads-start.

I would like to thank the companies ANSYS and ISimQ for providing me with example images that show the power of modern flow calculation programs. I would also like to thank Mr. Zipsner and the mechanical engineering editorial office of Springer Vieweg Verlag for their good support.

Bad Toelz, July 2017, Stefan Lecheler

Neubiberg, Germany   Stefan Lecheler

Symbol directory

Symbol	Unit	Designation
A	m^2	Area, surface
a	m/s^2	Acceleration
a	m/s	Speed of sound
CFL	–	Courant-Friedrichs-Levy Number
c_v	$J/kg/K$	Specific heat capacity at constant volume
c_p	$J/kg/K$	Specific heat capacity at constant pressure
dx,dy,dz	M	Side surfaces of the volume element dV
$\dot{E}$	Y/s	Energy flux
e	J/kg	Specific internal energy
ε	m^2/s^3	Turbulent dissipation
F	N	Force
F,G,H	–	Flow vectors in Cartesian coordinates
$\widehat{F},\widehat{G},\widehat{H}$	–	Flow vectors in curvilinear coordinates
g	m/s^2	Acceleration due to gravity
h	J/kg	Specific enthalpy
I	–	Unit matrix
$\dot{I}$	N/s	Momentum flux
i,j,k	–	Mesh point index in the three spatial directions
k	m^2/s^2	Turbulent kinetic energy
λ	$W/m/K$	Thermal conductivity coefficient
Ma	–	Mach number
m	kg	Mass
$\dot{m}$	kg/s	Mass flow
NBC	–	Numerical boundary conditions
$\vec{\triangledown}$	–	Divergence
n	–	Normal direction

(continued)

Symbol	Unit	Designation
μ	Pa s	Dynamic viscosity
ω	m^2/s^3	Turbulent frequency
PBC	–	Physical boundary conditions
p	Pa	Print
ϕ	–	Potential
Q	–	Source term in Cartesian coordinates
$\widehat{Q}$	–	Source term in curvilinear coordinates
$\dot{Q}$	W	Heat flux
$\dot{q}$	W/kg	Specific heat flux
R	J/kg/K	Gas constant
ρ	kg/m^3	Density
S	J/K	Entropy
T	K	Temperature
t,τ	s	Time
U	–	Conservation vector in Cartesian coordinates, also general flow quantity
$\widehat{U}$	–	Conservation vector in curvilinear coordinates
u,v,w	m/s	Velocity components in x-, y-, z-direction
V,dV	m^3	Volume, volume element
$\dot{W}$	W	Power
x,y,z	m	Cartesian spatial coordinates
ξ,η,ζ	–	Curvilinear spatial coordinates

Index	Designation
i,j,k	Spatial directions
x,y,z	x-, y-, z-direction
n	Time level
t	Total
w	Wall
$+$	Positive eigenvalue
$-$	Negative eigenvalue

Contents

Introduction 1

1.1 Aim of This Book

If you do not know anything about computational fluid dynamics (yet), this is the book for you. It is kept simple and understandable, at least from the author's point of view. It is aimed at prospective engineers in mechanical engineering and similar fields of study who are already familiar with fluid mechanics and thermodynamics. Knowledge of numerical solution methods and in computer science is an advantage, but not mandatory. The goals in detail are that you

- understand the theory behind CFD programs (CFD = Computational Fluid Dynamics),
- know the most important terms and equations from the CFD area,
- know how commercial CFD programs such as ANSYS CFX work and be able to operate them,
- be able to assess the scope, critical points and results of CFD programs.

This book does not aim to enable you to write your own CFD programs. This is usually not necessary when using commercial CFD programs. For readers who are interested in further details after reading this book, the review literature [1–5] and the technical literature [6–29] are available.

© The Author(s), under exclusive license to Springer Fachmedien Wiesbaden GmbH, part of Springer Nature 2022
S. Lecheler, *Computational Fluid Dynamics*,
https://doi.org/10.1007/978-3-658-38453-1_1

1.2   Tasks of the Numerical Flow Calculation

The flow equations (these are the conservation equations for mass, momentum and energy, as will be shown later) have theoretical solutions only for the simplest applications such as the flat plate or the cylindrical flow. Only for these special cases can the pressures, velocities and temperatures be calculated analytically. For real flow problems from, for example, aerospace, energy technology, automotive engineering, weapons technology, marine technology or medical technology, the flow must be determined either experimentally or numerically (Fig. 1.1).

In an experiment, a scale model of the body to be investigated is placed in the wind or water tunnel and the pressures, temperatures, velocities and forces are recorded using probes. Experiments are usually complex and expensive:

• The real Mach and Reynolds numbers often cannot be adjusted, which limits the similarity to the original. For example, high flow velocities as in hypersonic can only be achieved with great effort (by cryogenics) or for a short time (in the shock wave wind tunnel).
• Flow details can often not be recorded, such as the rotating internal flow in turbines, as probes can only be installed here with great effort for design reasons. Finally, many of the probes used also falsify the flow to be measured.

Numerical flow computation helps to overcome these disadvantages here. Through the development of powerful computers, the flow equations can be solved numerically and real problems from practice can be calculated accurately, quickly and cost-effectively. Particularly in the development of new products such as aircraft, rockets, automobiles and turbines, hundreds of variants have to be calculated for the fluidic design and optimization until, for example, the flow losses are minimal.

It must be pointed out, however, that the accuracy of the CFD program for the respective application area should always first be checked by means of a validation. Validation means that the calculation results are compared with theoretical and/or experimental results for

Fig. 1.1 The three disciplines of fluid mechanics

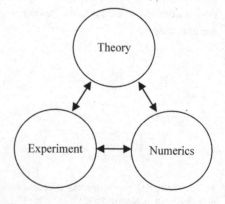

similar geometries and flows. Only when the agreement is satisfactory can CFD programs be reliably used for the design and optimization of new geometries.

Figures 1.2, 1.3, 1.4, 1.5, 1.6, 1.7, 1.8, 1.9, 1.10, 1.11 and 1.12 show typical examples of the performance of modern CFD programs:

- The calculation of the flow in a future piston engine, in which cryogenic hydrogen flows into the valve, can be seen in Fig. 1.2. The mass fraction of hydrogen at a crank angle of 460° is shown.

- Figure 1.3 shows the calculated streamlines around a Formula 1 racing car. The flow arrives uniformly, swirls at the wheels and detaches mainly behind the rear of the racing car.

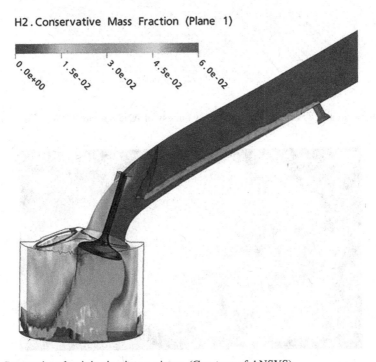

Fig. 1.2 Cryogenic valve injection into a piston. (Courtesy of ANSYS)

Fig. 1.3 Flow around a Formula 1 racing car. (Courtesy of ANSYS and BMW Sauber F1 Team)

Fig. 1.4 Air turbulence and aeroacoustic noise propagation around a passenger car. (Courtesy of ANSYS and FCA Italy)

- Meanwhile, modern CFD programs can also predict noise propagation due to air turbulence. Figure 1.4 shows these turbulences and the associated noise propagation around a passenger car.

- Figure 1.5 shows the results of a flow calculation around an aircraft wing with extended flaps. In addition to the streamlines, the turbulence intensity on the surface is also shown here. The color change shows the area of transition from laminar to turbulent flow.

- Figure 1.6 shows the result of a fluid-structure interaction calculation around a water turbine. In addition to the streamlines around the blades, the stresses in the rotor blades are also shown.

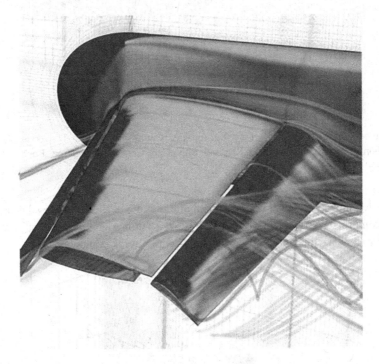

Fig. 1.5 Flow around an aircraft wing with flaps extended. (Courtesy of ANSYS)

Fig. 1.6 Flow and blade loading on a hydro turbine. (Courtesy of Cardiff University (CMERG), Low Carbon Research Institute (LCRI) and SuperGen Centre for Marine Energy Research (UKCMER))

- Figure 1.7 shows the Mach number distribution on the blades of a five-stage transonic axial compressor. Again, the red areas on the suction sides of the first and second impeller show a supersonic flow, which ends with a weak compression shock.

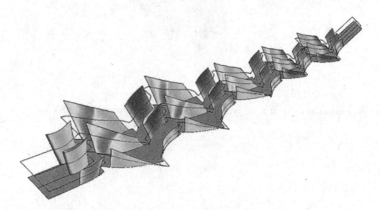

Fig. 1.7 Mach number distribution on the blades of a transonic axial compressor. (Courtesy of ALSTOM)

- Figure 1.8 shows the pressure distributions and the streamlines in a centrifugal compressor. In these calculations, the influence of the discharge diffuser on the pressure ratio and the efficiency was investigated.

- CFD calculations are also used in the field of sports to reduce resistance. Figure 1.9 shows the pressure distribution on a racing bike.

Fig. 1.8 Pressure distribution and streamlines in a centrifugal compressor. (Courtesy of ISimQ and PCA Engineering)

Fig. 1.9 Pressure distribution of a road bike. (Courtesy of ANSYS)

• Figure 1.10 shows the streamlines in the water when flowing around a swimmer. This
  allows the shape of the swimsuit to be optimised in terms of flow.

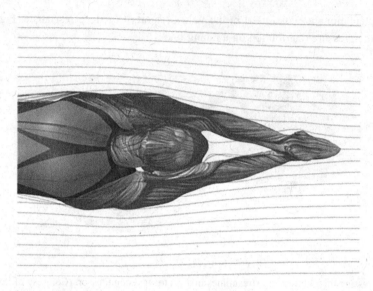

Fig. 1.10 Flow around a swimmer wearing a swimsuit. (Courtesy of Speedo and ANSYS)

• CFD methods are also used to calculate the flow in and around complex buildings.
  Figure 1.11 shows the streamlines around the Amsterdam football stadium.

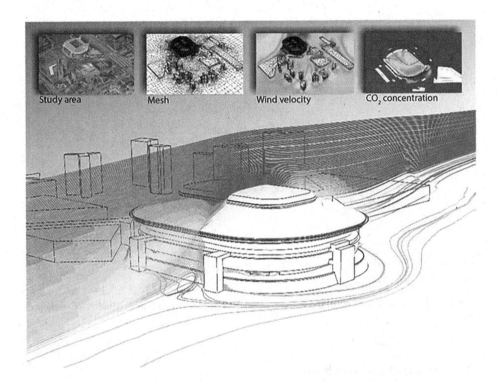

Study area   Mesh   Wind velocity CO_2 concentration

Fig. 1.11 Flow around the Amsterdam Arena football stadium. (Courtesy of Twan van Hoof and Bert Blocken, Eindhoven University of Technology)

- Figure 1.12 is entitled "Athena in the summer wind". For a Christmas card, the flow around the former logo of the University of the German Armed Forces was calculated. It shows the velocity vectors at an incident flow of 50 m/s. In terms of flow, it could certainly be optimized.

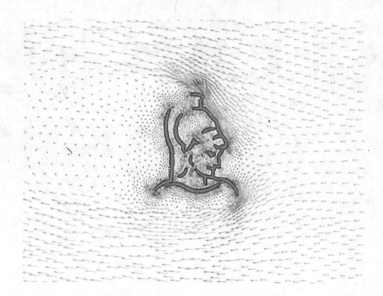

Fig. 1.12 Flow around the logo of the University of the Federal Armed Forces Munich. (Courtesy of UniBw Munich)

1.3   Structure of the Book

In Chap. 2 the flow equations are introduced. These are the **conservation equations for mass, momentum and energy**, as they are known from the lectures for fluid mechanics and thermodynamics. In addition to the **Navier-Stokes equations**, which include all physically relevant effects, common **simplification options** are also presented. These require shorter computation times with often sufficient accuracy.

Chapter 3 deals with the **discretization of the differential equations** presented in Chap. 2. This includes the formation of the spatial and the temporal differences, their order of accuracy and their influence on the stability.

Chapter 4 deals with the **different types of computational meshes** and their advantages and disadvantages. On the one hand, this includes the structured computational meshes in the form of rectangular, skew-angled and block-structured meshes. On the other hand, unstructured and adaptive computational meshes are also presented.

Chapter 5 deals with the solution methods of the systems of equations presented in Chaps. 2 and 3, respectively. In addition to the classical central methods, the **modern solution algorithms** of the upwind and high-resolution methods are presented, with which the flow quantities on the computational mesh points can be calculated accurately, quickly and robustly.

In Chap. 6, the **typical sequence of a numerical flow calculation** from the geometry generation to the evaluation and validation is exemplarily presented. The newcomer to the numerical flow calculation gets **numerous tips from practice**.

In Chaps. 7, 8 and 9, **three simple application examples** for the typical procedure of a numerical flow calculation with **ANSYS CFX** are presented. Chapter 7 deals with the flow around an **airfoil section**, Chap. 8 with the internal flow in a **pipe with an additional lateral inlet** and Chap. 9 with the heat transfer from a hot to a cold fluid in a **double-tube heat exchanger**. The CAD files for the three exercise examples can be downloaded from the homepage of the Laboratory of Thermodynamics of the Faculty of Mechanical Engineering of the University of the Federal Armed Forces Munich at www.unibw.de/mb/institute/we5/we51/downloads/downloads-start. At the beginning of each chapter, **objectives** are formulated in the form of questions. You should be able to answer these questions after reading the chapter. The correct **answers** can still be downloaded from the homepage of the Laboratory of Thermodynamics. However, they are now also additionally at the end of the book.

For feedback and questions, please also feel free to email me directly at stefan.lecheler@unibw.de.

Conservation Equations of Fluid Mechanics

2

2.1 Aim of This Chapter

I can still remember the beginning of my doctorate well. I took over a computer program from a mathematician for calculating flows in turbomachinery. While his description of the theory contained practically only one equation, the associated computer program was very extensive. How did that fit together? Well, mathematicians usually try to generalize everything and put it succinctly into a formula. It looks elegant, but an engineer will not understand its meaning in most cases, because he is not familiar with the special mathematical knowledge of vector and tensor calculus. After a few days and many pages of paper, I had transformed this one equation so that I understood it. My equations were longer, but more understandable to an engineer. This chapter therefore attempts to present the equations underlying fluid mechanics as simply as possible.

You should then be able to answer the following questions:

1. Which five quantities are retained in the flow calculation?
2. What is the difference between the integral and differential forms?
3. How does one arrive at the equation of conservation of mass?
4. From which equation can the equations of conservation of momentum be derived?
5. From which equation can the conservation of energy equation be derived?
6. What is the vectorial form of the Navier-Stokes equations with conservation vector, flux term and source term?
7. What other equations are needed to calculate the flow?
8. What is the difference between physical and numerical boundary conditions?
9. How many boundary conditions must be given at a subsonic inflow boundary, at a supersonic outflow boundary and at the solid boundary?

© The Author(s), under exclusive license to Springer Fachmedien Wiesbaden GmbH, part of Springer Nature 2022
S. Lecheler, *Computational Fluid Dynamics*,
https://doi.org/10.1007/978-3-658-38453-1_2

10. What is the difference between the complete and the Reynolds-averaged Navier-Stokes equations?
11. Why are turbulence models necessary for the latter?
12. What is neglected in the thin-layer Navier-Stokes equations?
13. Which terms are neglected in the Euler equations? For which Re numbers do they therefore only apply?
14. What simplifications lead to the potential equation?
15. Which equations are needed to calculate the polar (lift and drag coefficient versus angle of attack) of an airfoil?
16. You want to estimate the shock location around a hypersonic aircraft. Which equations would suffice?

2.2   Derivation of the Conservation Equations

Do conservation equations mean that they stay with us for a lifetime? At least for a fluid dynamics engineer, this is true. But the name comes, of course, from the conservation of certain physical quantities such as mass, momentum, and energy. This gives rise to the five conservation equations of fluid mechanics:

- Mass Conservation
- Conservation of momentum in *x-direction*
- Conservation of momentum in *y-direction*
- Conservation of momentum in *z-direction*
- Conservation of Energy

All modern CFD programs have in common that they solve these five conservation equations to calculate the flow of gases or liquids.

The conservation equations can be given in two different ways, in integral and in differential form. Table 2.1 shows the most important differences.

Both formulations can be transferred into each other. As a rule, the integral form or the finite volume discretization is used in modern CFD programs. It can better capture compaction shocks, since it allows for V discontinuities within the control volume, while the differential form or the finite difference discretization presupposes that the flow variables in the volume element V are differentiable, i.e. continuous. This is, however, not the case for compaction shocks.

In contrast, the differential form is mathematically more descriptive because it uses differentials. Therefore, for simplicity, the conservation equations are set up in differential form and the integral form is given only for comparison.

Table 2.1 Comparison of the integral and differential form of the conservation equations

	Integral form	Differential form
Fluid flows through	A finite control volume V	An infinitesimally small volume element V
Conservation equations	In integral form	In differential form
Discretization	As finite volume method (FV)	As finite difference (FD) method
Advantages	Physically more descriptive: Temporal change of the flow magnitude inside the control volume V corresponds to the change of the fluxes through the control surface A	Mathematically more descriptive, since no integrals appear
	Is more accurate for discontinuous courses such as for compaction joints	
Disadvantages	Mathematically more complex, because integrals appear	Physically more obscure, as volume approaches zero

2.2.1 Equation of Conservation of Mass

In a Cartesian coordinate system xy,z lies the solid volume element ($V = dx \cdot dy \cdot dz$ Fig. 2.1).

By this we mean:

$\rho = f(x, y, z, t)$	Density of the fluid as a function of the three spatial coordinates x, y, z and time t
$u, v, w = f(x, y, z, t)$	Flow velocities in x-, y and z direction
$m = \rho \cdot V = \rho \cdot dx \cdot dy \cdot dz$	Mass inside the volume element V

The mass balance is now established for this volume element:

- In the space-fixed volume element, V the change in mass with time corresponds to the partial derivative with respect to time $\frac{\partial}{\partial t}(m) = \frac{\partial}{\partial t}(\rho \cdot V) = \frac{\partial}{\partial t}(\rho \cdot dx \cdot dy \cdot dz)$.
- The mass flow entering $A_x = dy \cdot dz$ through the surface in the x-direction is $(\rho \cdot u) \cdot dy \cdot dz$ and the mass flow leaving is $\left[(\rho \cdot u) + \frac{\partial}{\partial x}(\rho \cdot u) \cdot dx\right] \cdot dy \cdot dz$.
- The mass flow entering $A_y = dx \cdot dz$ through the surface in the y-direction is $(\rho \cdot v) \cdot dx \cdot dz$ and the mass flow leaving is $\left[(\rho \cdot v) + \frac{\partial}{\partial y}(\rho \cdot v) \cdot dy\right] \cdot dx \cdot dz$.
- The mass flow entering $A_z = dx \cdot dy$ through the surface in the z-direction is $(\rho \cdot w) \cdot dx \cdot dy$ and the mass flow leaving is $\left[(\rho \cdot w) + \frac{\partial}{\partial z}(\rho \cdot w) \cdot dz\right] \cdot dx \cdot dy$.

For the mass balance then applies

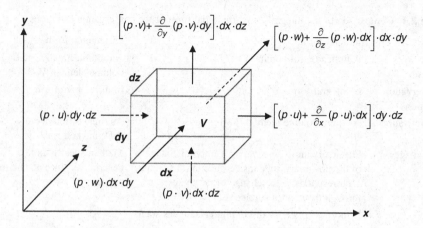

Fig. 2.1 Infinitesimal small volume element with mass flows

$$\frac{\partial}{\partial t}(\rho \cdot dx \cdot dy \cdot dz) + \left[(\rho \cdot u) + \frac{\partial}{\partial x}(\rho \cdot u) \cdot dx\right] \cdot dy \cdot dz - \rho \cdot u \cdot dy \cdot dz$$

$$+ \left[(\rho \cdot v) + \frac{\partial}{\partial y}(\rho \cdot v) \cdot dy\right] \cdot dx \cdot dz - \rho \cdot v \cdot dx \cdot dz$$

$$+ \left[(\rho \cdot w) + \frac{\partial}{\partial z}(\rho \cdot w) \cdot dz\right] \cdot dx \cdot dy - \rho \cdot w \cdot dx \cdot dy = 0.$$

After truncation of the $(\rho \cdot u)$, $(\rho \cdot v)$, $(\rho \cdot w)$-terms and division by, the **equation of conservation of mass in differential form in Cartesian coordinates** is obtained $dx \cdot dy \cdot dz$

$$\frac{\partial}{\partial t}(\rho) + \frac{\partial}{\partial x}(\rho \cdot u) + \frac{\partial}{\partial y}(\rho \cdot v) + \frac{\partial}{\partial z}(\rho \cdot w) = 0. \tag{2.1}$$

It means: the change of density ρ with time in the volume element plus the change of mass flow in $\rho \cdot u$ *x-direction plus the* change of mass flow in $\rho \cdot v$ *y-direction* plus the change of mass flow in $\rho \cdot w$ *z-direction* are zero.

For comparison, the **mass conservation equation in integral form** for the finite volume discretization **is** also given

$$\frac{\partial}{\partial t} \iiint\limits_{V} \rho \cdot dV + \iint\limits_{S} \rho \cdot \vec{u} \cdot dS = 0. \tag{2.2}$$

It means: the change of density ρ with time in the control volume V plus the change of mass flow $\rho \cdot \vec{u}$ over the surface of S the control volume is zero.

Equations 2.1 and 2.2 contain the same physics. This becomes clearer if one writes Eq. 2.1 as a **mass conservation equation in divergence form**

$$\frac{\partial}{\partial t}(\rho) + \vec{\nabla} \cdot \left(\rho \cdot \vec{u}\right) = 0 \tag{2.3}$$

with the divergence $\vec{\nabla} \cdot \left(\rho \cdot \vec{u}\right) = \frac{\partial}{\partial x}(\rho \cdot u) + \frac{\partial}{\partial y}(\rho \cdot v) + \frac{\partial}{\partial z}(\rho \cdot w)$ in Cartesian coordinates and the velocity vector in the $\vec{u} = u \cdot i + v \cdot j + w \cdot k$ three spatial directions i, j, k.

2.2.2 Conservation of Momentum Equations

The conservation of momentum is based on Newton's second law: force equals mass times acceleration $\vec{F} = m \cdot \vec{a}$ or with the momentum current $\vec{I} = \frac{d\vec{I}}{dt} = m \cdot \frac{d\vec{u}}{dt}$. For the three spatial directions then results

$$F_x = m \cdot a_x \quad \text{or} \quad \frac{dI_x}{dt} = m \cdot \frac{du}{dt}, \tag{2.4}$$

$$F_y = m \cdot a_y \quad \text{or} \quad \frac{dI_y}{dt} = m \cdot \frac{dv}{dt}, \tag{2.5}$$

$$F_z = m \cdot a_z \quad \text{or} \quad \frac{dI_z}{dt} = m \cdot \frac{dw}{dt}. \tag{2.6}$$

The force vector ($\vec{F}$, Fig. 2.2) includes both body forces such as gravity and the electromagnetic force and surface forces such as the compressive force and the frictional forces. The latter in turn consist of the normal stress force, which pulls the particle in length and the shear stress force, which shears the particle.

Figure 2.3 shows this time a Cartesian coordinate system x, y, z with an infinitesimally small volume element $V = dx \cdot dy \cdot dz$ and all the applied force components. The surfaces are again $A_x = dy \cdot dz$, and $A_y = dx \cdot dz A_z = dx \cdot dy$.

The sizes shown are:

$p = f(x, y, z, t)$	Fluid pressure
$\tau_{xx} = f(x, y, z, t)$	Normal stress in *x-direction* normal to $x = $ const. area $dy \cdot dz$
$\tau_{yx} = f(x, y, z, t)$	Shear stress in *x-direction* along the $y = $ const. surface $dx \cdot dz$
$\tau_{zx} = f(x, y, z, t)$	Shear stress in *x-direction* along the $z = $ const. surface $dx \cdot dy$

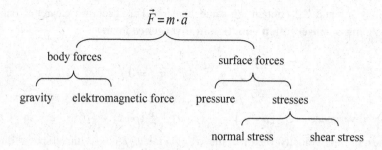

Fig. 2.2 Composition of the force for fluids

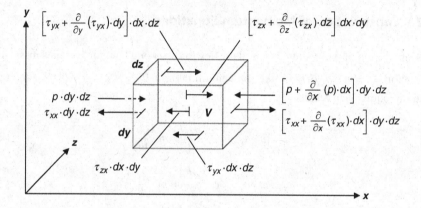

Fig. 2.3 The forces in the *x-direction* on an infinitesimally small volume element

In the following, the momentum equation in the *x-direction is* set up in differential form using the example of Eq. 2.4 and the respective more detailed terms are used for the force, the mass and the acceleration.

The force or the pulse current in the *x-direction is* thus given by

$$F_x = \frac{dI_x}{dt} = \left[\frac{\partial}{\partial x}(\tau_{xx}) + \frac{\partial}{\partial y}(\tau_{yx}) + \frac{\partial}{\partial z}(\tau_{zx}) - \frac{\partial p}{\partial x} + \rho \cdot g_x \right] \cdot dx \cdot dy \cdot dz. \qquad (2.7)$$

The mass inside the volume element is again, *V* as in the case of mass conservation

$$m = \rho \cdot V = \rho \cdot dx \cdot dy \cdot dz. \qquad (2.8)$$

The acceleration is the total derivative of the velocity with respect to time

$$a_x = \frac{du}{dt} = \frac{\partial u}{\partial t} + u \cdot \frac{\partial u}{\partial x} + v \cdot \frac{\partial u}{\partial y} + w \cdot \frac{\partial u}{\partial z}. \tag{2.9}$$

If Eqs. 2.7–2.9 are substituted into Eq. 2.4, it follows that

$$\underbrace{\left[\frac{\partial}{\partial x}(\tau_{xx}) + \frac{\partial}{\partial y}(\tau_{yx}) + \frac{\partial}{\partial z}(\tau_{zx}) - \frac{\partial p}{\partial x} + \rho \cdot g_x\right]}_{F_x} \cdot dx \cdot dy \cdot dz$$

$$= \underbrace{\rho \cdot dx \cdot dy \cdot dz}_{m} \cdot \underbrace{\left(\frac{\partial u}{\partial t} + u \cdot \frac{\partial u}{\partial x} + v \cdot \frac{\partial u}{\partial y} + w \cdot \frac{\partial u}{\partial z}\right)}_{a_x}$$

or

$$\frac{\partial}{\partial x}(\tau_{xx}) + \frac{\partial}{\partial y}(\tau_{yx}) + \frac{\partial}{\partial z}(\tau_{zx}) - \frac{\partial p}{\partial x} + \rho \cdot g_x = \rho \cdot \frac{\partial u}{\partial t} + \rho \cdot u \cdot \frac{\partial u}{\partial x} + \rho \cdot v \cdot \frac{\partial u}{\partial y} + \rho \cdot w$$

$$\cdot \frac{\partial u}{\partial z}.$$

This equation is the so-called non-conservative form of the *x-momentum equation,* since the terms ρ, $\rho \cdot u$, $\rho \cdot v$ and are $\rho \cdot w$ on the right-hand side before the derivatives. This non-conservative form of the conservation equation is less accurate in the later discretization, since the momentum is not completely conserved.

The conservative form does not have this disadvantage. With it, the momentum is completely preserved even after discretization. To get to the conservative form, the right side is further transformed so that all quantities are under the derivatives.

If the mathematical regularities for the transformation of differentials

$$\frac{\partial(\rho \cdot u)}{\partial t} = \rho \cdot \frac{\partial u}{\partial t} + u \cdot \frac{\partial \rho}{\partial t} \qquad \frac{\partial(\rho \cdot u \cdot v)}{\partial y} = \rho \cdot v \cdot \frac{\partial u}{\partial y} + u \cdot \frac{\partial(\rho \cdot v)}{\partial y}$$

$$\frac{\partial(\rho \cdot u^2)}{\partial x} = \rho \cdot u \cdot \frac{\partial u}{\partial x} + u \cdot \frac{\partial(\rho \cdot u)}{\partial x} \qquad \frac{\partial(\rho \cdot u \cdot w)}{\partial z} = \rho \cdot w \cdot \frac{\partial u}{\partial z} + u \cdot \frac{\partial(\rho \cdot w)}{\partial z}$$

resolved according to the non-conservative terms

$$\rho \cdot \frac{\partial u}{\partial t} = \frac{\partial(\rho \cdot u)}{\partial t} - u \cdot \frac{\partial \rho}{\partial t} \qquad \rho \cdot v \cdot \frac{\partial u}{\partial y} = \frac{\partial(\rho \cdot u \cdot v)}{\partial y} - u \cdot \frac{\partial(\rho \cdot v)}{\partial y}$$

$$\rho \cdot u \cdot \frac{\partial u}{\partial x} = \frac{\partial(\rho \cdot u^2)}{\partial x} - u \cdot \frac{\partial(\rho \cdot u)}{\partial x} \qquad \rho \cdot w \cdot \frac{\partial w}{\partial z} = \frac{\partial(\rho \cdot u \cdot w)}{\partial z} - u \cdot \frac{\partial(\rho \cdot w)}{\partial z},$$

follows for the right side

$$\rho \cdot \frac{\partial u}{\partial t} + \rho \cdot u \cdot \frac{\partial u}{\partial x} + \rho \cdot v \cdot \frac{\partial u}{\partial y} + \rho \cdot w \cdot \frac{\partial u}{\partial z}$$

$$= \frac{\partial(\rho \cdot u)}{\partial t} - u \cdot \frac{\partial \rho}{\partial t} + \frac{\partial(\rho \cdot u^2)}{\partial x} - u \cdot \frac{\partial(\rho \cdot u)}{\partial x} + \frac{\partial(\rho \cdot u \cdot v)}{\partial y} - u \cdot \frac{\partial(\rho \cdot v)}{\partial y}$$

$$+ \frac{\partial(\rho \cdot u \cdot w)}{\partial z} - u \cdot \frac{\partial(\rho \cdot w)}{\partial z}$$

$$= \frac{\partial(\rho \cdot u)}{\partial t} + \frac{\partial(\rho \cdot u^2)}{\partial x} + \frac{\partial(\rho \cdot u \cdot v)}{\partial y} + \frac{\partial(\rho \cdot u \cdot w)}{\partial z}$$

$$- u \cdot \underbrace{\left[\frac{\partial \rho}{\partial t} + \frac{\partial(\rho \cdot u)}{\partial x} + \frac{\partial(\rho \cdot v)}{\partial y} + \frac{\partial(\rho \cdot w)}{\partial z} \right]}_{\text{Mass Conservation} = 0}$$

$$= \frac{\partial(\rho \cdot u)}{\partial t} + \frac{\partial(\rho \cdot u^2)}{\partial x} + \frac{\partial(\rho \cdot u \cdot v)}{\partial y} + \frac{\partial(\rho \cdot u \cdot w)}{\partial z}.$$

The term in the square bracket disappears because it is identical to the equation of conservation of mass according to Eq. 2.1 and thus becomes zero.

The x-momentum conservation equation then follows in **differential form in Cartesian coordinates**

$$\frac{\partial}{\partial t}(\rho \cdot u) + \frac{\partial}{\partial x}\left(\rho \cdot u^2 + p - \tau_{xx}\right) + \frac{\partial}{\partial y}\left(\rho \cdot u \cdot v - \tau_{yx}\right)$$

$$+ \frac{\partial}{\partial z}(\rho \cdot u \cdot w - \tau_{zx}) - \rho \cdot g_x = 0. \tag{2.10}$$

Similarly, the **y-momentum conservation equation in differential form in Cartesian coordinates is** given by Eq. 2.5

$$\frac{\partial}{\partial t}(\rho \cdot v) + \frac{\partial}{\partial x}\left(\rho \cdot v \cdot u - \tau_{xy}\right) + \frac{\partial}{\partial y}\left(\rho \cdot v^2 + p - \tau_{yy}\right) \backslash$$

$$+ \frac{\partial}{\partial z}\left(\rho \cdot v \cdot w - \tau_{zy}\right) - \rho \cdot g_y = 0 \tag{2.11}$$

and the **z-momentum conservation equation in differential form in Cartesian coordinates** from Eq. 2.6

$$\frac{\partial}{\partial t}(\rho \cdot w) + \frac{\partial}{\partial x}\left(\rho \cdot w \cdot u - \tau_{xz}\right) + \frac{\partial}{\partial y}\left(\rho \cdot w \cdot v - \tau_{yz}\right)$$

$$+ \frac{\partial}{\partial z}\left(\rho \cdot w^2 + p - \tau_{zz}\right) - \rho \cdot g_z = 0. \tag{2.12}$$

Note: The frictional normal and shear stresses τ can still be expressed by the velocity gradients. These expressions are different for the so-called "Newtonian fluids" and the so-called "non-Newtonian fluids":

- For Newtonian fluids, according to Newton, the stresses are proportional to the velocity gradients. This model is practically always valid in aerodynamics.
- For non-Newtonian fluids, the stresses are not proportional to the velocity gradient. This is the case for highly viscous fluids such as honey.

2.2.3 Conservation of Energy Equation

The conservation of energy is known as the first law of thermodynamics

$$\frac{dE_{\text{ges}}}{dt} = \dot{W} + \dot{Q} \tag{2.13}$$

and means that the change in total energy E_{ges} in the volume element is equal to the power $\dot{W}$ at the volume element plus the heat flow $\dot{Q}$ into the volume element. Equation 2.13 applies to a closed system without mass transfer, which is why the volume element must move with the flow here. The three terms total energy, power and heat flow are described in more detail below.

The total energy is composed E_{ges} three parts:

- of the internal energy $E_{\text{in}} = m \cdot e$ with the specific internal energy e,
- of the kinetic energy with $E_{\text{kin}} = \frac{1}{2} \cdot m \cdot \vec{u}^2 = \frac{1}{2} \cdot m \cdot (u^2 + v^2 + w^2) \; \vec{u}^2 = u^2 + v^2 + w^2$ as the square of the velocity magnitude
- and the potential energy $E_{\text{pot}} = m \cdot g \cdot h$, which can be neglected for gases and is therefore neglected for the sake of clarity.

Thus, the total energy of a gas is as follows

$$E_{\text{ges}} = E_{\text{in}} + E_{\text{kin}} = m \cdot e + \frac{1}{2} \cdot m \cdot \vec{u}^2 = m \cdot \left(e + \frac{1}{2} \cdot \vec{u}^2\right) = \rho \cdot V \cdot \left(e + \frac{1}{2} \cdot \vec{u}^2\right)$$

$$= \rho \cdot dx \cdot dy \cdot dz \cdot \left(e + \frac{1}{2} \cdot \vec{u}^2\right) = \rho \cdot \left(e + \frac{1}{2} \cdot \vec{u}^2\right) \cdot dx \cdot dy \cdot dz$$

and the total derivative with respect to time is

$$\frac{dE_{\text{ges}}}{dt} = \frac{d}{dt}\left[\rho \cdot \left(e + \frac{1}{2} \cdot \vec{u}^2\right)\right] \cdot dx \cdot dy \cdot dz = \frac{\partial}{\partial t}\left[\rho \cdot \left(e + \frac{1}{2} \vec{u}^2\right)\right]$$

$$+ \frac{\partial}{\partial x}\left[\rho \cdot u \cdot \left(e + \frac{1}{2} \cdot \vec{u}^2\right)\right] + \frac{\partial}{\partial y}\left[\rho \cdot v \cdot \left(e + \frac{1}{2} \cdot \vec{u}^2\right)\right] \tag{2.14}$$

$$+ \frac{\partial}{\partial z}\left[\rho \cdot w \cdot \left(e + \frac{1}{2} \cdot \vec{u}^2\right)\right] \cdot dx \cdot dy \cdot dz.$$

Power $\dot{W}$ is the power due to body and surface forces and is the product of force times velocity component in the direction of force. The power is composed of the following three parts:

- of gravity g, which acts on the volume element,
- the pressure p acting on the surfaces of the volume element
- and the normal and shear stresses τ which also act on the surfaces.

Using the flows in the *x-direction* from Fig. 2.4 and the corresponding flows in the y- and z-directions, the power is given by

$$\dot{W} = \left[\rho \cdot (u \cdot g_x + v \cdot g_y + w \cdot g_z) - \frac{\partial}{\partial x}(u \cdot p) - \frac{\partial}{\partial y}(v \cdot p) - \frac{\partial}{\partial z}(w \cdot p)\right. \tag{2.15}$$

$$+ \frac{\partial}{\partial x}\left(u \cdot \tau_{xx} + v \cdot \tau_{xy} + w \cdot \tau_{xz}\right) + \frac{\partial}{\partial y}\left(u \cdot \tau_{yx} + v \cdot \tau_{yy} + w \cdot \tau_{yz}\right)$$

$$+ \left. \frac{\partial}{\partial z}\left(u \cdot \tau_{zx} + v \cdot \tau_{zy} + w \cdot \tau_{zz}\right)\right] dx \cdot dy \cdot dz.$$

The heat flow is $\dot{Q}$ again composed of three parts:

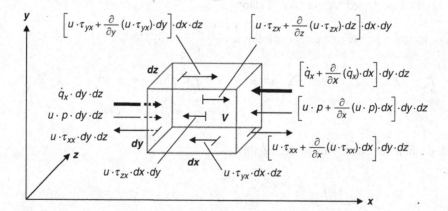

Fig. 2.4 The power and heat flows in the *x-direction* at an infinitesimally small volume element

- the heat conduction $\dot{q}_L$ over the surface of the volume element, caused by temperature gradients
- the thermal radiation $\dot{q}_S$ acting on the volume element (neglected here for the sake of clarity)
- and convection, which however does not appear here, because the volume element swims along with the flow.

This results in the following for the heat flow

$$\dot{Q} = \left[-\frac{\partial}{\partial x}(\dot{q}_{L,x}) - \frac{\partial}{\partial y}(\dot{q}_{L,y}) - \frac{\partial}{\partial z}(\dot{q}_{L,z}) \right] \cdot dx \cdot dy \cdot dz.$$

Using Fourier's laws of heat conduction $\dot{q}_{L,x} = -\lambda \cdot \frac{\partial T}{\partial x}$, and $\dot{q}_{L,y} = -\lambda \cdot \frac{\partial T}{\partial y}$ $\dot{q}_{L,z} = -\lambda \cdot \frac{\partial T}{\partial z}$ gives the following equation.

$$\dot{Q} = \left[\frac{\partial}{\partial x}\left(\lambda \cdot \frac{\partial T}{\partial x}\right) + \frac{\partial}{\partial y}\left(\lambda \cdot \frac{\partial T}{\partial y}\right) + \frac{\partial}{\partial z}\left(\lambda \cdot \frac{\partial T}{\partial z}\right) \right] \cdot dx \cdot dy \cdot dz. \tag{2.16}$$

If Eqs. 2.14–2.16 are inserted into Eq. 2.13, the energy conservation equation follows with to $\vec{u}^2 = u^2 + v^2 + w^2$

$$\frac{\partial}{\partial t}\left[\rho \cdot \left(e + \frac{1}{2}\cdot \vec{u}^2\right) \right]$$

$$+ \frac{\partial}{\partial x}\left[\rho \cdot u \cdot \left(e + \frac{1}{2}\cdot \vec{u}^2\right) \right] + \frac{\partial}{\partial y}\left[\rho \cdot v \cdot \left(e + \frac{1}{2}\cdot \vec{u}^2\right) \right] + \frac{\partial}{\partial z}\left[\rho \cdot w \cdot \left(e + \frac{1}{2}\cdot \vec{u}^2\right) \right]$$

$$- \rho \cdot (u \cdot g_x + v \cdot g_y + w \cdot g_z) + \frac{\partial}{\partial x}(u \cdot p) + \frac{\partial}{\partial y}(v \cdot p) + \frac{\partial}{\partial z}(w \cdot p)$$

$$- \frac{\partial}{\partial x}(u \cdot \tau_{xx} + v \cdot \tau_{xy} + w \cdot \tau_{xz}) - \frac{\partial}{\partial y}(u \cdot \tau_{yx} + v \cdot \tau_{yy} + w \cdot \tau_{yz})$$

$$- \frac{\partial}{\partial z}(u \cdot \tau_{zx} + v \cdot \tau_{zy} + w \cdot \tau_{zz}) - \frac{\partial}{\partial x}\left(\lambda \cdot \frac{\partial T}{\partial x}\right) - \frac{\partial}{\partial y}\left(\lambda \cdot \frac{\partial T}{\partial y}\right)$$

$$- \frac{\partial}{\partial z}\left(\lambda \cdot \frac{\partial T}{\partial z}\right) = 0.$$

This equation can be further simplified. If the specific enthalpy is used $h = e + p/\rho$ instead of the specific internal energy e for the spatial derivatives, the pressure term is omitted.

If the terms are then rearranged, $\partial/\partial x$, $\partial/\partial y$, $\partial/\partial z$ **the equation of conservation of energy is obtained in differential form in Cartesian coordinates**

$$\frac{\partial}{\partial t}\left[\rho \cdot \left(e + \frac{1}{2} \cdot \vec{u}^2\right)\right] + \frac{\partial}{\partial x}\left[\rho \cdot u \cdot \left(h + \frac{1}{2} \cdot \vec{u}^2\right) - \left(u \cdot \tau_{xx} + v \cdot \tau_{xy} + w \cdot \tau_{xz}\right) - \lambda \cdot \frac{\partial T}{\partial x}\right]$$

$$+ \frac{\partial}{\partial y}\left[\rho \cdot v \cdot \left(h + \frac{1}{2} \cdot \vec{u}^2\right) - \left(u \cdot \tau_{yx} + v \cdot \tau_{yy} + w \cdot \tau_{yz}\right) - \lambda \cdot \frac{\partial T}{\partial y}\right]$$

$$+ \frac{\partial}{\partial z}\left[\rho \cdot w \cdot \left(h + \frac{1}{2} \cdot \vec{u}^2\right) - \left(u \cdot \tau_{zx} + v \cdot \tau_{zy} + w \cdot \tau_{zz}\right) - \lambda \cdot \frac{\partial T}{\partial z}\right]$$

$$- \rho \cdot \left(u \cdot g_x + v \cdot g_y + w \cdot g_z\right) = 0.$$

(2.17)

2.3   Navier-Stokes Equations

2.3.1   Complete Navier-Stokes Equations

The five conservation equations of mass, momentum and energy are called the Navier-Stokes equations (in the past, only the three momentum conservation equations were often called this). Complete means that the flow is described completely, i.e. also with the smallest vortices and turbulences. However, the computation times for the solution of the complete Navier-Stokes equations are so large that for technical applications the so-called Reynolds-averaged Navier-Stokes equations are used, as will be shown later in Sect. 2.3.4.

The Navier-Stokes equations describe the motion of a viscous, isotropic fluid and are so named in honour of Messrs. Navier and Stokes. Their most important life data can be seen in Fig. 2.5.

Claude Louis Marie Henri Navier
- French physicist and engineer
- 15 February 1785 to 23 August 1836
- From 1819 Professor in Paris
- Significant contributions to mechanics, structural analysis and hydromechanics

Sir George Gabriel Stokes
- British mathematician and physicist
- 13 August 1819 to 1 February 1903
- From 1849 Professor at Cambridge
- 1885–90 President of the Royal Society
- Important contributions to analysis (Stokes integral theorem), optics (Stokes rule) and hydrodynamics (Stokes friction law)

Fig. 2.5 The namesakes of the Navier-Stokes equations

The Navier-Stokes equations form a coupled nonlinear differential equation system and could so far be solved analytically only for special cases such as one-dimensional flows around the plane plate. Therefore, this system of equations is solved numerically for general cases.

The Navier-Stokes equations can be represented in different ways:

- scalar
- in vector
- in divergence form.

These three forms are described in more detail below.

Navier-Stokes Equations in Scalar Form (Cartesian Coordinates)

The scalar form is the form used in the last chapter. For the sake of clarity, the five conservation equations for mass (Eq. 2.1), momentum (Eqs. 2.10–2.12) and energy (Eq. 2.17) are given again

$$\frac{\partial}{\partial t}(\rho) + \frac{\partial}{\partial x}(\rho \cdot u) + \frac{\partial}{\partial y}(\rho \cdot v) + \frac{\partial}{\partial z}(\rho \cdot w) = 0$$

$$\frac{\partial}{\partial t}(\rho \cdot u) + \frac{\partial}{\partial x}(\rho \cdot u^2 + p - \tau_{xx})$$
$$+ \frac{\partial}{\partial y}(\rho \cdot u \cdot v - \tau_{yx}) + \frac{\partial}{\partial z}(\rho \cdot u \cdot w - \tau_{zx}) - \rho \cdot g_x = 0$$

$$\frac{\partial}{\partial t}(\rho \cdot v) + \frac{\partial}{\partial x}(\rho \cdot v \cdot u - \tau_{xy})$$
$$+ \frac{\partial}{\partial y}(\rho \cdot v^2 + p - \tau_{yy}) + \frac{\partial}{\partial z}(\rho \cdot v \cdot w - \tau_{zy}) - \rho \cdot g_y = 0$$

$$\frac{\partial}{\partial t}(\rho \cdot w) + \frac{\partial}{\partial x}(\rho \cdot w \cdot u - \tau_{xz}) + \frac{\partial}{\partial y}(\rho \cdot w \cdot v - \tau_{yz})$$
$$+ \frac{\partial}{\partial z}(\rho \cdot w^2 + p - \tau_{zz}) - \rho \cdot g_z = 0$$

$$+ \frac{\partial}{\partial t}\left[\rho \cdot \left(e + \frac{1}{2} \cdot \vec{u}^2\right)\right] + \frac{\partial}{\partial x}\left[\rho \cdot u \cdot \left(h + \frac{1}{2} \cdot \vec{u}^2\right) - (u \cdot \tau_{xx} + v \cdot \tau_{xy} + w \cdot \tau_{xz}) - \lambda \cdot \frac{\partial T}{\partial x}\right]$$
$$+ \frac{\partial}{\partial y}\left[\rho \cdot v \cdot \left(h + \frac{1}{2} \cdot \vec{u}^2\right) - (u \cdot \tau_{yx} + v \cdot \tau_{yy} + w \cdot \tau_{yz}) - \lambda \cdot \frac{\partial T}{\partial y}\right]$$
$$+ \frac{\partial}{\partial z}\left[\rho \cdot w \cdot \left(h + \frac{1}{2} \cdot \vec{u}^2\right) - (u \cdot \tau_{zx} + v \cdot \tau_{zy} + w \cdot \tau_{zz}) - \lambda \cdot \frac{\partial T}{\partial z}\right]$$
$$- \rho \cdot (u \cdot g_x + v \cdot g_y + w \cdot g_z) = 0$$

The Navier-Stokes equations contain three types of terms:

- The time derivative of $\frac{\partial}{\partial t}$ the conservation variables $\vec{U}$ with the density ρ, momentum $\rho \cdot \vec{u}$ and total energy $\rho \cdot \left(e + \frac{1}{2} \cdot \vec{u}^2 \right)$. They give the changes of the sought variables in the volume element with time. According to them, the equations are solved.

- The spatial derivatives $\frac{\partial}{\partial x}, \frac{\partial}{\partial y}, \frac{\partial}{\partial z}$ of the flow terms $\vec{E}$, and $\vec{F}$ $\vec{G}$ with the convective terms with $\rho \cdot \vec{u}$, the pressure terms with p, the friction terms with the stresses τ and the heat conduction terms with λ. They indicate what flows into and out of the volume element through all surfaces.

- The terms without derivatives are called $\vec{Q}$ source terms, because they act like a source in the volume element. This is the gravity with the gravitational constant g (and possibly the thermal radiation $\dot{q}_S$).

Navier-Stokes Equations in Vector Form (Cartesian Coordinates)
More elegantly and briefly, the Navier-Stokes equations can be written in vector form. It is also useful for the implementation in a computer program

$$\frac{\partial}{\partial t} \vec{U} + \frac{\partial}{\partial x} \vec{E} + \frac{\partial}{\partial y} \vec{F} + \frac{\partial}{\partial z} \vec{G} = \vec{Q} \tag{2.18}$$

with the conservation vector $\vec{U}$

$$\vec{U} = \begin{bmatrix} \rho \\ \rho \cdot u \\ \rho \cdot v \\ \rho \cdot w \\ \rho \cdot \left[e + \frac{1}{2} \left(u^2 + v^2 + w^2 \right) \right] \end{bmatrix}, \tag{2.19}$$

the flow vectors in x $\vec{E}, \vec{F}, \vec{G}$-, y and z *direction*

$$\vec{E} = \begin{bmatrix} \rho \cdot u \\ \rho \cdot u^2 + p - \tau_{xx} \\ \rho \cdot v \cdot u - \tau_{xy} \\ \rho \cdot w \cdot u - \tau_{xz} \\ \rho \cdot u \cdot \left[h + \frac{1}{2} \left(u^2 + v^2 + w^2 \right) \right] - u \cdot \tau_{xx} - v \cdot \tau_{xy} - w \cdot \tau_{xz} - \lambda \cdot \frac{\partial T}{\partial x} \end{bmatrix}, \tag{2.20}$$

$$\vec{F} = \begin{bmatrix} \rho \cdot v \\ \rho \cdot u \cdot v - \tau_{yx} \\ \rho \cdot v^2 + p - \tau_{yy} \\ \rho \cdot w \cdot v - \tau_{yz} \\ \rho \cdot v \cdot \left[h + \frac{1}{2}\left(u^2 + v^2 + w^2\right)\right] - u \cdot \tau_{yx} - v \cdot \tau_{yy} - w \cdot \tau_{yz} - \lambda \cdot \dfrac{\partial T}{\partial y} \end{bmatrix}, \quad (2.21)$$

$$\vec{G} = \begin{bmatrix} \rho \cdot w \\ \rho \cdot u \cdot w - \tau_{zx} \\ \rho \cdot v \cdot w - \tau_{zy} \\ \rho \cdot w^2 + p - \tau_{zz} \\ \rho \cdot w \cdot \left[h + \frac{1}{2}\left(u^2 + v^2 + w^2\right)\right] - u \cdot \tau_{zx} - v \cdot \tau_{zy} - w \cdot \tau_{zz} - \lambda \cdot \dfrac{\partial T}{\partial z} \end{bmatrix} \quad (2.22)$$

and the so-called source term $\vec{Q}$

$$\vec{Q} = \begin{bmatrix} 0 \\ \rho \cdot g_x \\ \rho \cdot g_y \\ \rho \cdot g_z \\ \rho \cdot \left(u \cdot g_x + v \cdot g_y + w \cdot g_z\right) \end{bmatrix}. \quad (2.23)$$

Navier-Stokes Equations in Divergence Form

A more mathematical notation, but independent of the coordinate system, can also often be found in the literature. It uses the velocity vector $\vec{U}$, the gravity vector $\vec{g}$, the divergence $\vec{\nabla}$, the unit matrix I and the stress matrix

$$\vec{u} = \begin{bmatrix} u \\ v \\ w \end{bmatrix} \quad \vec{g} = \begin{bmatrix} g_x \\ g_y \\ g_z \end{bmatrix} \quad \vec{\nabla} = \begin{bmatrix} \dfrac{\partial}{\partial x} \\ \dfrac{\partial}{\partial y} \\ \dfrac{\partial}{\partial z} \end{bmatrix} \quad I = \begin{bmatrix} 1 & 0 & 0 \\ 0 & 1 & 0 \\ 0 & 0 & 1 \end{bmatrix} \quad \tau = \begin{bmatrix} \tau_{xx} & \tau_{xy} & \tau_{xz} \\ \tau_{yx} & \tau_{yy} & \tau_{yz} \\ \tau_{zx} & \tau_{zy} & \tau_{zz} \end{bmatrix}.$$

Hereby the equation of conservation of mass becomes

$$\frac{\partial}{\partial t}(\rho) + \vec{\nabla} \cdot \left(\rho \cdot \vec{u}\right) = 0$$

and the three equations of conservation of momentum with the vector product are $\otimes$

$$\frac{\partial}{\partial t}\left(\rho \cdot \vec{u}\right) + \vec{\nabla} \cdot \left(\rho \cdot \vec{u} \otimes \vec{u} + p \cdot I - \tau\right) = \rho \cdot \vec{g}$$

and the equation of conservation of energy is given by

$$\frac{\partial}{\partial t}\left[\rho \cdot \left(e + \frac{1}{2} \cdot \vec{u}^2\right)\right] + \vec{\nabla} \cdot \left[\rho \cdot \vec{u} \cdot \left(h + \frac{1}{2} \cdot \vec{u}^2\right) - \tau \cdot \vec{u} - \lambda \cdot \vec{\nabla} T\right] = \rho \cdot \vec{g} \cdot \vec{u}.$$

Summarized again to a system of equations we get

$$\frac{\partial}{\partial t}\begin{bmatrix} \rho \\ \rho \cdot \vec{u} \\ \rho \cdot \left(e + \frac{1}{2} \cdot \vec{u}^2\right) \end{bmatrix} + \vec{\nabla}\begin{bmatrix} \rho \cdot \vec{u} \\ \rho \cdot \vec{u} \otimes \vec{u} + p \cdot I - \tau \\ \rho \cdot \vec{u} \cdot \left(h + \frac{1}{2} \cdot \vec{u}^2\right) - \tau \cdot \vec{u} - \lambda \cdot \vec{\nabla} T \end{bmatrix} = \begin{bmatrix} 0 \\ \rho \cdot \vec{g} \\ \rho \cdot \vec{g} \cdot \vec{u} \end{bmatrix}$$

or abbreviated

$$\frac{\partial}{\partial t}\vec{U} + \vec{\nabla} \cdot \vec{F} = \vec{Q}. \tag{2.24}$$

With this formulation in divergence form the analogy to the integral form becomes visible

$$\frac{\partial}{\partial t}\iiint_V \vec{U} \cdot dV + \iint_S \vec{F} \cdot d\vec{S} = \iiint_V \vec{Q} \cdot dV. \tag{2.25}$$

2.3.2 Additionally Required Equations and Quantities

The five conservation equations are not sufficient for the solution, because they contain more unknowns than equations. In order to determine all 17 unknowns (ρ, u, v, w, p, e, h, T and the 9τ terms), 12 further equations are required in addition to the five conservation equations. These are three equations of state for the fluid and nine so-called Stokes relations for the normal and shear stresses.

The **thermal equation of state** couples the pressure p with the density ρ and the temperature T. For example, for an ideal gas with R as gas constant it is

$$p = \rho \cdot R \cdot T. \tag{2.26}$$

The **caloric equations of state** couple the specific internal energy e and the specific enthalpy h with the temperature T. For an ideal gas it reads with c_v as specific heat capacity at constant volume and c_p as specific heat capacity at constant pressure e.g.

$$de = c_v \cdot dT, \tag{2.27}$$

$$dh = c_p \cdot dT. \tag{2.28}$$

The **Stokes relations** couple the stresses with τ velocities for the so-called Newtonian fluids u, v, w. The terms are with μ as dynamic viscosity of the fluid

$$\tau_{xx} = -\frac{2}{3}\mu \cdot \left(\frac{\partial u}{\partial x} + \frac{\partial v}{\partial y} + \frac{\partial w}{\partial z}\right) + 2 \cdot \mu \cdot \frac{\partial u}{\partial x}, \tag{2.29}$$

$$\tau_{yy} = -\frac{2}{3}\mu \cdot \left(\frac{\partial u}{\partial x} + \frac{\partial v}{\partial y} + \frac{\partial w}{\partial z}\right) + 2 \cdot \mu \cdot \frac{\partial v}{\partial y}, \tag{2.30}$$

$$\tau_{zz} = -\frac{2}{3}\mu \cdot \left(\frac{\partial u}{\partial x} + \frac{\partial v}{\partial y} + \frac{\partial w}{\partial z}\right) + 2 \cdot \mu \cdot \frac{\partial w}{\partial z}, \tag{2.31}$$

$$\tau_{xy} = \mu \cdot \left(\frac{\partial v}{\partial x} + \frac{\partial u}{\partial y}\right), \tag{2.32}$$

$$\tau_{xz} = \mu \cdot \left(\frac{\partial u}{\partial z} + \frac{\partial w}{\partial x}\right), \tag{2.33}$$

$$\tau_{yz} = \mu \cdot \left(\frac{\partial w}{\partial y} + \frac{\partial v}{\partial z}\right), \tag{2.34}$$

$$\tau_{yx} = \tau_{xy} = \mu \cdot \left(\frac{\partial v}{\partial x} + \frac{\partial u}{\partial y}\right), \tag{2.35}$$

$$\tau_{zx} = \tau_{xz} = \mu \cdot \left(\frac{\partial u}{\partial z} + \frac{\partial w}{\partial x}\right), \tag{2.36}$$

$$\tau_{zy} = \tau_{yz} = \mu \cdot \left(\frac{\partial w}{\partial y} + \frac{\partial v}{\partial z}\right). \tag{2.37}$$

2.3.3   The Substance Values

It should also be noted that the substance values depend on the temperature, as is usually the case: c_v, c_p, λ, μ

- The specific heat capacities c_v, c_p and the coefficient of thermal conductivity λ are often assumed to be constant or interpolated from tables.
- The dynamic viscosity μ can also be interpolated from tables. For air, it is often also calculated approximately using the so-called Sutherland formula

$$\mu = 1{,}458 \cdot 10^{-6} \cdot \left(\frac{T^{1,5}}{T + 110,4} \right) \qquad \text{with } T \text{ in [K]}. \qquad (2.38)$$

Boundary Conditions

The Navier-Stokes equations shown above apply to (almost) all flows. Both the supersonic flow around an airplane and the subsonic flow around a car or also the internal flow in a pipe can be calculated with it. How is a distinction made between these different flow cases, when the equations are the same? In order to be able to solve the concrete flow problem, boundary conditions are therefore necessary in addition to the respective geometry. Typical boundary conditions are, for example:

- What flows into the calculation area (inflow edge)?
- What flows out of the calculation area (outflow edge)?
- What is the flow like at a solid wall (solid edge)?

Only by correct choice of boundary conditions also a flow is established. For example, the total pressure in the inflow must be higher than in the outflow.

Since the five unknowns remain ρ, u, v, w, e in the five conservation equations, five quantities must also be known at the boundary. Depending on the type of boundary and the flow, however, not all five quantities may be given, some quantities must also be calculated:

- The specified quantities at the boundary are called **physical boundary conditions (PBC)**. They are specified by the user in the form of known quantities, usually from measurements or theoretical derivations. These are usually pressures, velocities, temperatures or mass flows. Their distribution at the boundary can be constant or variable.
- the calculable quantities at the boundary are called **numerical boundary conditions (NBC)**. These are equations that link the boundary to the inner field and are calculated by the program. The most accurate way is to use selected conservation equations, since these correctly represent the physics. Less accurate but simpler is to extrapolate the flow quantities from the internal field to the boundaries.

The sum of PBC and NBC must always correspond to the number of conservation equations to be solved. Thus, for the number of physical and numerical boundary conditions, depending on the spatial dimension, the following result

$$3D: \quad PBC + NBC = 5,$$
$$2D: \quad PBC + NBC = 4, \qquad\qquad (2.39)$$
$$1D: \quad PBC + NBC = 3.$$

The number of physical boundary conditions PBC can be determined from the so-called characteristics theory. This theory describes, among other things, the direction of propagation of information in a flow. For example, in a supersonic flow there is no upstream effect; all characteristics travel downstream. This leads to the fact that in a supersonic airplane, the sonic boom propagates only to the rear; nothing is heard from the front. In a subsonic flow, on the other hand, one characteristic runs upstream and transports information against the direction of flow. This can be seen, for example, in a ship. Most water waves run backwards, but a bow wave also spreads out in front of the ship.

As a rule, therefore:

- Physical boundary conditions must be specified for all information or characteristics that come to the computational domain from outside.
- Numerical boundary conditions are used for all information or characteristics that run from the inside to the edge of the computational domain.

The number of physical boundary conditions therefore depends on how the flow crosses the boundary (inflow-, outflow-, solid boundary) and whether there is subsonic or supersonic flow at the boundary.

Physical Boundary Conditions at the Inflow Edge

For a subsonic inflow, four physical boundary conditions must be specified according to Fig. 2.6 (this applies to 3D problems, for 2D problems it is three and for 1D problems two). As a rule, the following four quantities are specified for a subsonic inflow (Table 2.2):

- the total pressure p_t
- the total temperature T_t
- the inflow direction in the *xz-plane* (inflow angle γ) or the ratio of the inflow velocity w/u
- and the inflow direction in the *xy-plane* (inflow angle α) or the ratio of the inflow velocity v/u.

Depending on the program, other variables can also be selected, such as the mass flow. However, the above four have proven to be particularly robust. Their values can be

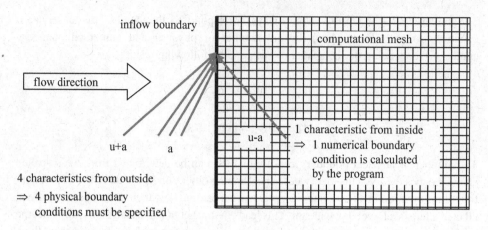

inflow boundary

computational mesh

flow direction

u+a   a

u-a

1 characteristic from inside
⇒ 1 numerical boundary
condition is calculated
by the program

4 characteristics from outside
⇒ 4 physical boundary
conditions must be specified

Fig. 2.6 Characteristics at the subsonic inflow edge

Table 2.2 Number and usual variables (*in brackets*) *of* physical boundary conditions (PBC) and number of numerical boundary conditions (NBC)

Dimension	Margins PBC	Subsonic NBC		PBC	Supersonic NBC	
3D	Inflow edge	$4\left(p_t, T_t, \frac{v}{u}, \frac{w}{u}\right)$	1	$5\,(p_t, T_t, u, v, w)$	0
	Outflow edge	$1\,(p)$	4	0	5
	Solid state edge	$4\,(u, v, w, T)$	1	$4\,(u, v, w, T)$	1
2D	Inflow edge	$3\left(p_t, T_t, \frac{v}{u}\right)$	1	$4\,(p_t, T_t, u, v)$	0
	Outflow edge	$1\,(p)$	3	0	4
	Solid state edge	$3\,(u, v, T)$	1	$3\,(u, v, T)$	1
1D	Inflow edge	$2\,(p_t, T_t)$	1	$3\,(p_t, T_t, u)$	0
	Outflow edge	$1\,(p)$	2	0	3
	Solid state edge	$2\,(u, T)$	1	$2\,(u, T)$	1

constant or variable. For example, a measured boundary layer profile for total pressure and temperature can also be specified.

For a supersonic inflow, all five physical boundary conditions must be specified (Fig. 2.7).

In most cases, the following quantities are selected for a supersonic inflow (Table 2.2):

- the total pressure p_t
- the total temperature T_t
- the inflow velocity in *x-direction* u
- the inflow velocity in *y-direction* v
- the inflow velocity in the *z-direction* w.

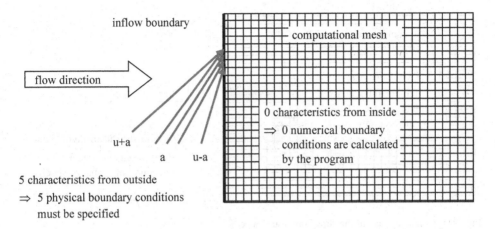

Fig. 2.7 Characteristics at the supersonic inflow edge

Physical Boundary Conditions at the Downstream Edge

At the downstream boundary, a distinction is also made between subsonic and supersonic flow. In subsonic, only one physical boundary condition is required (Fig. 2.8). Mostly the static pressure $p = $ const. or a static pressure distribution is given. $p = f(y, z)$.

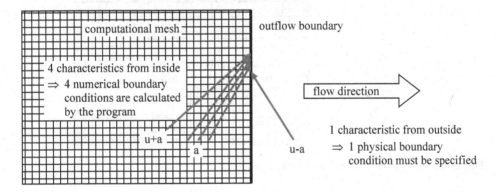

Fig. 2.8 Characteristics at the subsonic outflow edge

In supersonic, on the other hand, no physical boundary condition needs to be specified at all (Fig. 2.9).

Physical Boundary Conditions at the Solid State Boundary

At the edge of the solid, both in subsonic (Fig. 2.10) and in supersonic (Fig. 2.11), four quantities must be specified as physical boundary conditions. These are the three velocity components and the wall temperature or its gradient.

For the velocity components, the so-called **no-slip condition** applies to flows with friction, i.e. the velocities at the wall are zero.

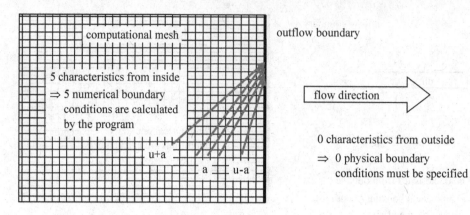

Fig. 2.9 Characteristics at the supersonic outflow edge

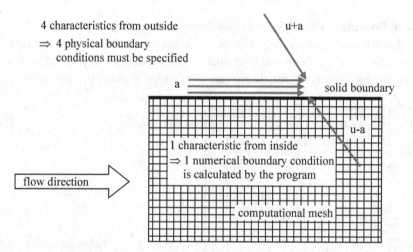

Fig. 2.10 Characteristics at the subsonic solid edge

$$u = v = w = 0. \tag{2.40}$$

In most CFD programs, the velocity components do not have to be explicitly specified, but the option "frictional wall" is selected.

There are three formulations for the wall temperature, depending on whether the wall temperature is known or whether the wall is thermally permeable or not.

If the wall temperature is T_w known, the following applies

$$T = T_w. \tag{2.41}$$

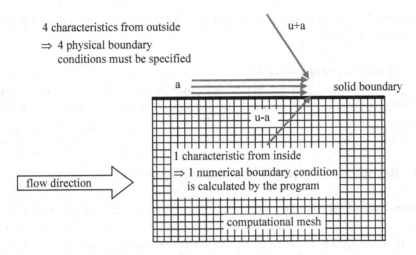

Fig. 2.11 Characteristics at the supersonic solid edge

If the heat flux is known, the temperature gradient normal to the wall can be given. From Fourier's law of heat conduction follows

$$\left(\frac{\partial T}{\partial n}\right)_w = -\frac{\dot{q}_w}{\lambda}. \tag{2.42}$$

If the wall is heat-impermeable (adiabatic), the temperature gradient normal to the wall is as follows

$$\left(\frac{\partial T}{\partial n}\right)_w = 0. \tag{2.43}$$

Non-Reflective Boundary Conditions

If constant values are specified for the physical boundary conditions at the inflow and outflow edges, these edges must be far enough away from the body (e.g. airfoil, turbine blade, aircraft, vehicle) so as not to disturb the flow at the body itself. However, this is not always possible. For example, when calculating the flow through blade rows of turbomachines, the distances between the guide vane and the impeller are usually so small that constant boundary conditions between the guide vane and the impeller can influence and falsify the pressure distribution on the blade.

Here, a non-reflecting formulation of the inflow and outflow boundary conditions provides more accurate results. Non-reflective means that disturbances or sound waves at the inflow or outflow boundary of the computational domain are not reflected, but leave the computational domain almost undisturbed. Another advantage is that the computational

domain can be kept significantly smaller, which shortens the computation time or increases
the accuracy.

Summary of Boundary Conditions

Table 2.2 summarizes again the number of physical (PBC) and numerical boundary
conditions (NBC) and shows the usual quantities given as physical boundary conditions
in brackets.

2.3.4   Reynolds-Averaged Navier-Stokes Equations

The three-dimensional, so-called "complete" Navier-Stokes equations given at the beginning of Sect. 2.3 can be solved numerically. However, for turbulent flows the computational effort is very high, since due to the nonlinearity even the smallest disturbances
influence the solution. This means that even the smallest turbulences must still be resolved.
The volume elements must therefore be very small or the computational mesh must have a
very large number of points. This makes the computation times for technical applications
unacceptably large.

Therefore, today's CFD programs are based on the Reynolds-averaged Navier-Stokes
equations, which still represent the physics accurately enough, but lead to acceptable
computation times. With the Reynolds-averaged Navier-Stokes equations, the small
turbulences are not resolved, but are modeled by so-called turbulence models. The
differences are summarized in Fig. 2.12.

Complete Navier-Stokes-Equations

- All frequencies are included
- Turbulence is calculated directly
- Mesh must be very fine
- Computing times are very long

Reynolds-averaged Navier-Stokes-Equations

- Only medium and low frequencies are included
- High frequencies are calculated by turbulence models
- Mesh can be coarser because high frequencies must not be resolved
- Computing times are acceptable

Fig. 2.12 Complete and Reynolds-averaged Navier-Stokes equations

To obtain the Reynolds-averaged Navier-Stokes equations, the individual flow quantities are replaced ρ', u', v', w', $e'\rho$, u, v, w, e by the sum of their low-frequency mean values and $\bar{\rho},\bar{u},\bar{v},\bar{w},\bar{e}$ their high-frequency fluctuation quantities, as in the following example

$$
\begin{aligned}
\rho &= \bar{\rho} + \rho', \\
u &= \bar{u} + u', \\
v &= \bar{v} + v', \\
w &= \bar{w} + w', e = \bar{e} + e'.
\end{aligned}
\tag{2.44}
$$

The high-frequency fluctuation terms capture the turbulent fluctuations in the flow and lead to the so-called Reynolds stresses. They are replaced by turbulence models. This has the advantage that the computational mesh no longer has to resolve the small turbulent fluctuations.

The Reynolds-averaged Navier-Stokes equations then only contain the low-frequency mean values. With them, the "normal" temporal fluctuations of the flow can still be captured, but not the turbulence. The Reynolds-averaged Navier-Stokes equations look formally like the full Navier-Stokes equations. However, there would be a transverse bar above each flow quantity, but again this is omitted for simplicity. The details are rather mathematical and for those interested can be found in the relevant literature such as [1–5].

2.3.5 Turbulence Models

Through this Reynolds averaging of the Navier-Stokes equations, however, the accuracy of the entire solution depends on the turbulence model. Above all, the position of the flow transition from laminar to turbulent and the location of the flow separation in the case of a pressure increase are not yet calculated accurately enough by many turbulence models.

The range of turbulence models used today extends from simple algebraic equations to second order differential equation systems:

- The laminar model for purely laminar flows, in which the Reynolds stresses disappear
- The eddy viscosity models, which replace the Reynolds stresses with a turbulent viscosity or eddy viscosity:
  - in the zero-equation model, the eddy viscosity is approximated by a simple algebraic equation and no differential equation for the eddy transport is solved. A typical representative is the Baldwin-Lomax model.
  - In the single-equation model, the eddy viscosity is calculated with a differential equation for the transport of the turbulent kinetic energy. A typical representative is the Spalart-Allmaras model.

– In the two-equation model, the eddy viscosity is determined from two differential equations. These include the k-ε-, k-ω and SST (Shear Stress Transport) models currently commonly used for industrial applications.

• The Reynolds stress models, which calculate the individual components of the Reynolds stress tensor and take into account the directional dependence of the turbulence.

• The eddy simulation models, which dispense with Reynolds averaging and solve the full Navier-Stokes equations transiently, such as the LES (Large Eddy Simulation), DES (Detached Eddy Simulation) and DNS (Direct Numerical Simulation) models.

The accuracy of the numerical solution increases with increasing complexity of the turbulence models. However, the computation time also increases strongly, especially if they require very fine computational meshes. The turbulence models commonly used in practice are described in some more detail below. Further details can be found in [6].

The **laminar model** is used for purely laminar flows. Here, the direct solution of the Navier-Stokes equations is sufficienti.e. no turbulence model is required. The condition is that the Reynolds numbers are small, otherwise an unphysical solution is obtained with the laminar model.

The **Baldwin-Lomax model** was originally developed for airfoil flows and is simple and robust. It is called a zero-equation model because it uses only one algebraic equation and no transport equations. The disadvantage of this model is its inaccuracy for detached flows. The detachment is calculated too late and the size of the detachment zone is too small. It was developed for structured computational meshes with orthogonal mesh lines on the wall, which is why a structured background mesh is additionally required for unstructured meshes.

The **k-ε model** uses two additional transport equations for turbulent kinetic energy k and turbulent dissipation ε. It is stable, requires little additional computational effort, and has long been the industry standard. It is well suited for the calculation of flow inside the flow field, but has problems in the calculation of flows that detach due to pressure gradients on the wall. The onset of detachment is calculated too late and the detachment area is calculated too small. The k-ε-model therefore usually delivers too optimistic results for flows that detach at the wall.

The **k-ω model** as the second representative of the two-equation models provides more accurate results than the k-ε model near the wall. This is achieved by using the turbulent ω frequency instead of the ε turbulent dissipation. It provides more accurate results for dissipated flows even at lower boundary layer resolution. In the interior of the flow field, however, it is inferior to the k-ε model with respect to accuracy.

For this reason, the so-called **SST model** (shear stress transport) was developed. It combines the good properties of the k-ω-model near the wall with the good properties of the k-ε-model in the remaining flow field. Thus, it provides more accurate results in the entire flow field, even for detached flows. Moreover, it is robust and the computation times are acceptable. The SST model is considered as the new standard turbulence model for industrial applications and also provides good results for phenomena such as pressure-induced detachment or heat transfer.

The **Reynolds stress model** should be used if the turbulence is anisotropic, i.e. direction-dependent, or non-equilibrium effects occur. In this model, instead of the isotropic eddy viscosity, the Reynolds stresses are calculated directly algebraically or via transport equation models. It is most accurate for complex flows with strong secondary flows, but the calculation times are significantly higher than for the eddy viscosity models.

Finally, the following points should be noted for all turbulence models:

- For all differential equations in the turbulence model, physical boundary conditions must also be specified.
- The location of the transition from laminar to turbulent flow is normally calculated automatically by a transition model. If it is known, e.g. from measurements, it can also be specified, which can improve the accuracy.

2.4 Simplification Possibilities

2.4.1 Introduction

Most commercial CFD programs are based on the Reynolds-averaged Navier-Stokes equations. The computation times are typically hours or days. Since many variants often have to be calculated in a short time when designing new geometries, short calculation times are very helpful. Therefore, CFD programs based on simplified conservation equations are often used in industry, as long as the physics is still adequately represented. For example, the pressure distributions around profiles with high Reynolds numbers can also be calculated relatively accurately without friction using the Euler equations. If, in addition, no compaction impacts occur, the potential equations are also sufficient. In this case, typical calculation times are seconds or minutes. Figure 2.13 shows the most important simplification options of the Navier-Stokes equations used today.

In the simplification, a distinction is made primarily according to the influence of viscosity. If the areas in which viscous effects such as friction and heat conduction occur are small, then certain friction terms can be omitted from the Reynolds-averaged Navier-Stokes equations.

2.4.2 Thin-Layer Navier-Stokes Equations

For the Thin-Layer-Navier-Stokes equations, the viscous terms and heat conduction terms tangential to the wall are neglected. Only the viscous terms and the heat conduction terms normal to the wall are retained. This has the advantage that the most important friction terms are still considered when the boundary layer is in contact, but the computation times

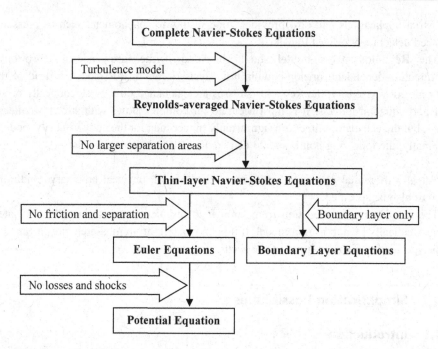

Fig. 2.13 Some simplification possibilities of the Navier-Stokes equations

are shorter than for the Reynolds-averaged Navier-Stokes equations. The short computation times result mainly from the fact that the computational mesh can be made coarser in the two spatial directions in which the viscous terms are neglected.

They can be used to calculate lift and drag. Their use makes sense if the flow has high Reynolds numbers and no large separation areas, such as in the case of airfoil flow at angles of attack around the design point.

The Thin-Layer-Navier-Stokes equations result from the Reynolds-averaged or the complete Navier-Stokes equations by neglecting the viscous terms with τ and the heat conduction terms with λ tangential to the wall in the flux vectors. For a wall area with $z = $ constant, the flux terms $\vec{E}$ and $\vec{F}$ change from Eqs. 2.20 and 2.21 to:

$$\vec{E} = \begin{bmatrix} \rho \cdot u \\ \rho \cdot u^2 + p \\ \rho \cdot v \cdot u \\ \rho \cdot w \cdot u \\ \rho \cdot u \cdot \left[h + \frac{1}{2} \left(u^2 + v^2 + w^2 \right) \right] \end{bmatrix}, \tag{2.45}$$

$$\vec{F} = \begin{bmatrix} \rho \cdot v \\ \rho \cdot u \cdot v \\ \rho \cdot v^2 + p \\ \rho \cdot w \cdot v \\ \rho \cdot v \cdot \left[h + \frac{1}{2} \left(u^2 + v^2 + w^2 \right) \right] \end{bmatrix}, \tag{2.46}$$

$$\vec{G} = \begin{bmatrix} \rho \cdot w \\ \rho \cdot u \cdot w - \tau_{zx} \\ \rho \cdot v \cdot w - \tau_{zy} \\ \rho \cdot w^2 + p - \tau_{zz} \\ \rho \cdot w \cdot \left[h + \frac{1}{2} \left(u^2 + v^2 + w^2 \right) \right] - u \cdot \tau_{zx} - v \cdot \tau_{zy} - w \cdot \tau_{zz} - \lambda \cdot \frac{\partial T}{\partial z} \end{bmatrix}$$

(as Eq. 2.22),

while the flux term $\vec{G}$ normal to the wall (Eq. 2.22) remains unchanged.
The tensions from the Stokes relations of Eqs. 2.32–2.37 then shorten to

$$\tau_{zx} = \mu \cdot \left(\frac{\partial u}{\partial z} \right), \tag{2.47}$$

$$\tau_{zy} = \mu \cdot \left(\frac{\partial v}{\partial z} \right), \tag{2.48}$$

$$\tau_{zz} = \frac{4}{3} \mu \cdot \left(\frac{\partial w}{\partial z} \right). \tag{2.49}$$

2.4.3   Euler Equations

The Euler equations are obtained from the Navier-Stokes equations neglecting all friction terms with τ and all heat conduction terms with λ. The flow terms Eqs. 2.20–2.22 then become

$$\vec{E} = \begin{bmatrix} \rho \cdot u \\ \rho \cdot u^2 + p \\ \rho \cdot v \cdot u \\ \rho \cdot w \cdot u \\ \rho \cdot u \cdot \left[h + \frac{1}{2} \left(u^2 + v^2 + w^2 \right) \right] \end{bmatrix}$$

(as in Eq. 2.45),

$$\vec{F} = \begin{bmatrix} \rho \cdot v \\ \rho \cdot u \cdot v \\ \rho \cdot v^2 + p \\ \rho \cdot w \cdot v \\ \rho \cdot v \cdot \left[h + \frac{1}{2} \left(u^2 + v^2 + w^2 \right) \right] \end{bmatrix}$$

(as in Eq. 2.46),

$$\vec{G} = \begin{bmatrix} \rho \cdot w \\ \rho \cdot u \cdot w \\ \rho \cdot v \cdot w \\ \rho \cdot w^2 + p \\ \rho \cdot w \cdot \left[h + \frac{1}{2} \left(u^2 + v^2 + w^2 \right) \right] \end{bmatrix}. \tag{2.50}$$

Since the Euler equations do not calculate friction in the boundary layer and thus do not have to resolve it, coarser calculation meshes are also sufficient. This allows the computation times to be significantly reduced. This is especially important for the design of new geometries. Here, many different geometries have to be examined until an optimum is found.

Since the Euler equations assume frictionless flow, the physical solid boundary condition also changes. Instead of the no-slip condition (Eq. 2.40 $u = v = w = 0$), the so-called slip condition now applies. The velocity vector must be tangential to the wall or its normal component must disappear.

$$u_n = \vec{u} \cdot \vec{n} = u \cdot n_x + v \cdot n_y + w \cdot n_z = 0. \tag{2.51}$$

The model of the Euler equations is applicable to flows at high Reynolds numbers and without detachments. An important property of the Euler equations (like the Navier-Stokes equations) is that they can automatically capture discontinuities such as compression shocks, since they satisfy the Rankine-Hugoniot conditions. The propagation of compression shocks in supersonic and hypersonic flows can thus already be captured relatively accurately with the Euler equations. If, for example, the friction losses are also to be calculated, they can also be coupled with the boundary layer equations.

2.4.4 Boundary Layer Equations

The so-called boundary layer equations are based on the investigations of Prandtl, who found that at high Reynolds numbers the viscous regions extend mainly along the solid walls and, provided no detachment occurs, the remaining regions are almost frictionless.

The boundary layer equations are the simplified momentum equations along the wall, which result from the Thin-Layer-Navier-Stokes equations

$$\frac{\partial}{\partial t}(\rho \cdot u) + \frac{\partial}{\partial x}(\rho \cdot u^2) + \frac{\partial}{\partial y}(\rho \cdot u \cdot v) + \frac{\partial}{\partial z}(\rho \cdot u \cdot w) + \frac{\partial p}{\partial x} - \frac{\partial}{\partial z}(\tau_{zx}) = 0, \quad (2.52)$$

$$\frac{\partial}{\partial t}(\rho \cdot v) + \frac{\partial}{\partial x}(\rho \cdot v \cdot u) + \frac{\partial}{\partial y}(\rho \cdot v^2) + \frac{\partial}{\partial z}(\rho \cdot v \cdot w) + \frac{\partial p}{\partial y} - \frac{\partial}{\partial z}(\tau_{zy}) = 0, \quad (2.53)$$

$$\frac{\partial p}{\partial z} = 0. \quad (2.54)$$

It can be seen that the *z-momentum equation* reduces to the pressure term normal to the wall, i.e. the pressure in the boundary layer remains constant. This decouples the pressure field from the viscous velocity field and can be calculated separately.

The practical procedure, e.g. around an airfoil, is then as follows:

- Frictionless calculation of the flow e.g. by means of the Euler equations or the potential equation.
- Transfer of the static pressure field at the wall to the boundary layer method.
- Calculation of velocities and losses in the boundary layer using the boundary layer method. Determination of the displacement thickness.
- Thickening of the airfoil by the displacement thickness.
- Repeat the frictionless calculation around the thickened profile.

The procedure is repeated until the solutions from the frictionless calculation of the core flow and the frictioned calculation of the boundary layer flow do not change any more or only minimally.

2.4.5 Potential Equation

As a last simplification possibility, the potential equation is given here. It is a further simplification of the Euler equations and is therefore only valid for frictionless flows. In order to arrive at the potential form, the flow must also be rotation-free Rotation-free means physically that both the entropy and the total enthalpy must be constant along streamlines (isentropic and isenthalpic flow). This is the case when no discontinuities such as collisions occur in the flow field. The potential equations cannot capture stronger compression shocks and are therefore only used for pure subsonic flows and transonic flows up to $Ma < 1.2$, where no or only weak shocks occur.

If the rotational freedom applies, τ the scalar potential can be introduced instead of the three velocities φ

$$\vec{u} = \vec{\nabla}\varphi \quad \text{or} \quad \begin{bmatrix} u \\ v \\ w \end{bmatrix} = \begin{bmatrix} \dfrac{\partial \varphi}{\partial x} \\ \dfrac{\partial \varphi}{\partial y} \\ \dfrac{\partial \varphi}{\partial z} \end{bmatrix}. \tag{2.55}$$

This reduces the five Euler equations tosingle potential equation

$$\frac{\partial}{\partial t}(\rho) + \vec{\nabla} \cdot \left(\rho \cdot \vec{\nabla}\varphi\right) = 0 \quad \text{or}$$
$$\frac{\partial}{\partial t}(\rho) + \frac{\partial}{\partial x}\left[\rho \cdot \frac{\partial \varphi}{\partial x}\right] + \frac{\partial}{\partial y}\left[\rho \cdot \frac{\partial \varphi}{\partial y}\right] + \frac{\partial}{\partial z}\left[\rho \cdot \frac{\partial \varphi}{\partial z}\right] = 0. \tag{2.56}$$

With the potential equation, airfoil or blade profiles can be designed very quickly. Even coupled with a boundary layer method, calculation times of only a few seconds result. They are used in industry, for example, for the optimization of so-called supercritical airfoil profiles, which are used in modern commercial aircraft in order to be able to fly shock-free and thus with low losses in the transonic range. They are also used for the design of subsonic and transonic compressor blades.

Finally, Fig. 2.14 shows an overview of the most important properties of the Navier-Stokes equations, Euler equations and the potential equation.

Navier-Stokes Equations

- Coupled system of 5 non-linear differential equations of 2nd order in space and time
- Fulfill conservation of mass, momentum and energy
- Describe wave propagation (convection), damped by friction

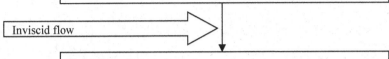

Inviscid flow

Euler Equations

- Coupled system of 5 non -linear differential equations of 1st order in space and time
- Fulfill conservation of mass, momentum and energy
- Describe wave propagation (convection) without friction

Non-rotational and isentropic flow

Potential Equation

- Non-linear differential equation of 2nd order in space and time
- Fulfills conservation of mass and energy
- Conservation of momentum is only fulfilled in absence of discontinuities like compression shocks

Fig. 2.14 Properties of the Navier-Stokes and Euler equations and the potential equation

Discretization of the Conservation Equations

3

3.1 Aim of This Chapter

The conservation equations shown in Chap. 2 have to be transformed so that they can be solved numerically in a computer program. This so-called discretization of the spatial and temporal derivatives is shown in this chapter. The user of modern flow calculation programs should at least have an overview of the discretization methods, even if these steps have already been taken from him by the developers of the CFD programs.

After reading this chapter, he should then be able to answer the following questions:

1. What is meant by discretization?
2. What are the three discretization methods?
3. What are the discretization options for the spatial derivatives?
4. What is meant by first and second order accuracy?
5. Which spatial order should be used for continuous flow and which for discontinuous flow?
6. How can the time derivative be discretized?
7. What does time asymptotic and time accurate computation mean?
8. Which time discretization is used for a transient calculation?
9. What is the advantage of solving the time derivatives for stationary solutions as well?
10. What are the names of the equations that result after discretization?
11. What is consistency?
12. When is a numerical solution procedure stable?
13. When is the numerical solution convergent?
14. Why is additive numerical viscosity needed for central spatial discretization?
15. What is meant by upwind discretization?

S. Lecheler, *Computational Fluid Dynamics*, https://doi.org/10.1007/978-3-658-38453-1_3

16. What is the difference between explicit and implicit discretization?
17. What are the advantages and disadvantages of implicit discretization?
18. Which number couples the time step with the mesh size?
19. How large can this number be in an explicit procedure?

3.2   What Does Discretization Mean?

For the basic equations of fluid mechanics given in Chap. 2 (Navier-Stokes equations or mass, − momentum and energy conservation equations) no analytical solutions are known, except for special cases like the plane plate. An analytical solution would be, for example, an equation for the density ρ as a function of the other quantities u, v, w, e.

Therefore, the differential equation system must be solved numerically for technically relevant problems. For this purpose, the partial derivatives (differentials) must be converted into finite differences. This is called discretization. The discretized differential equations are then called difference equations. These difference equations can be solved on a so-called computational mesh. In the numerical solution, the numerical values for the flow variables are then available as ρ, u, v, w, e at the mesh points.

Figure 3.1 shows a small section of a two-dimensional computational mesh with nine mesh points. The solution is calculated only at these grid points. A so-called cell corner point scheme is shown here, where the grid points are located on the corners of the volume element. Alternatively, they can also be located inside the volume element. The volume element is then offset by half a mesh size and is referred to as a cell center method.

Discretization thus means that in the differential equations the differentials are replaced by differences. For example, the differential of the velocity in $\partial u/\partial x$ *x-direction* at the point can be $P(i,j)$ replaced by the difference of the values at the neighboring points and $(i+1,j)$ (i,j)

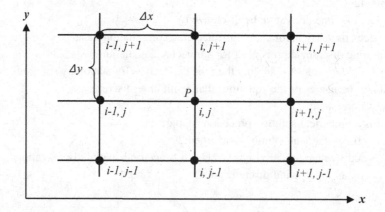

Fig. 3.1 Schematic of a computational mesh around the point P

$$\frac{\partial u}{\partial x} \approx \frac{\Delta u}{\Delta x} = \frac{u_{i+1,j} - u_{i,j}}{x_{i+1,j} - x_{i,j}} = \frac{u_{i+1,j} - u_{i,j}}{\Delta x}.$$

Three methods of discretization are distinguished, but they are equivalent and can be transferred into each other: the finite difference, the finite volume and the finite element discretization. Each method has its advantages and disadvantages:

- The **finite difference (FD)** discretization is very illustrative and is used below to show the principle of discretization. It uses the conservation equations in differential form (Chap. 2). The grid points are located at the corners of the volume element (cell corner point scheme).
- **Finite volume (FV) discretization** is very accurate for discontinuities such as shocks, which is why modern CFD programs mostly use this method. It uses the conservation equations in integral form and the integrals are replaced by sums. The support points are either located inside the volume element (cell center scheme) or, as in the FD discretization, at the corners (cell corner point scheme).
- The **finite element (FE) method** is preferred by mathematicians because it is mathematically easy to represent. In this method, mathematical equations such as straight line or parabolic equations are used for the differentials.

Figure 3.2 shows that finite difference (FD) methods have the highest accuracy and finite element (FE) methods have the highest flexibility. In practice, the finite volume (FV) methods have prevailed in the commercial CFD programs, since they have a good accuracy and flexibility.

In the following, the discretization of the differential equations mentioned in Chap. 2 is described in more detail. For reasons of clarity, this is done using the finite difference

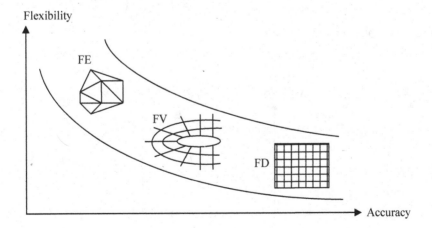

Fig. 3.2 Classification of discretization methods. (According to Laurien and Oertel [4])

method as an example. Since the discretization methods are different, a distinction is made between the discretization of the spatial derivatives and the temporal derivative.

3.3   Spatial Discretization

3.3.1   Discretization of the First Derivatives

The spatial discretization concerns the transformation of the partial derivatives (differentials)$\partial/\partial x$, and $\partial/\partial y\partial/\partial z$ into finite differences $\Delta/\Delta x$, and $\Delta/\Delta y\Delta/\Delta z$. The example of the partial spatial derivative is $\partial U/\partial x$ used to show the transformation possibilities into finite differences, where here is U representative of any flow quantity.

Looking at Fig. 3.3, there are three ways to form this derivative at the point P:

the forward differential	$\left(\frac{\partial U}{\partial x}\right)_{i,j} \approx \frac{U_{i+1,j}-U_{i,j}}{x_{i+1,j}-x_{i,j}} = \frac{U_{i+1,j}-U_{i,j}}{\Delta x}$
The backward difference	$\left(\frac{\partial U}{\partial x}\right)_{i,j} \approx \frac{U_{i,j}-U_{i-1,j}}{x_{i,j}-x_{i-1,j}} = \frac{U_{i,j}-U_{i-1,j}}{\Delta x}$
The central difference	$\left(\frac{\partial U}{\partial x}\right)_{i,j} \approx \frac{U_{i+1,j}-U_{i-1,j}}{x_{i+1,j}-x_{i-1,j}} = \frac{U_{i+1,j}-U_{i-1,j}}{2\Delta x}$

The question now is, what is the difference and what difference should be used? The difference arises when an attempt is made to replace the approximate sign with an equal sign. For this, the so-called truncation error must be introduced, which says with which order of magnitude the difference approximates the differential i.e. approximates.

For this purpose the classical Taylor series expansion is used

$$U_{i+1,j} = U_{i,j} + (\Delta x) \cdot \left(\frac{\partial U}{\partial x}\right)_{i,j} + \frac{(\Delta x)^2}{2} \cdot \left(\frac{\partial^2 U}{\partial x^2}\right)_{i,j} + \frac{(\Delta x)^3}{6} \cdot \left(\frac{\partial^3 U}{\partial x^3}\right)_{i,j} + \dots \quad (3.1)$$

Solved for the first derivative results in

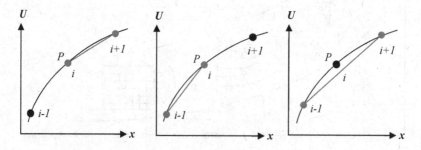

Fig. 3.3 Forward, backward and central spatial difference

$$\left(\frac{\partial U}{\partial x}\right)_{i,j} = \underbrace{\frac{U_{i+1,j} - U_{i,j}}{\Delta x}}_{\text{Finite Difference}} \underbrace{- \frac{(\Delta x)}{2} \cdot \left(\frac{\partial^2 U}{\partial x^2}\right)_{i,j} - \frac{(\Delta x)^2}{6} \cdot \left(\frac{\partial^3 U}{\partial x^3}\right)_{i,j} - \cdots}_{\text{Termination error}}$$

The terms of higher order are neglected and called truncation error. From this now results the discretization of the differential by the **forward difference of first order**

$$\left(\frac{\partial U}{\partial x}\right)_{i,j} = \frac{U_{i+1,j} - U_{i,j}}{\Delta x} + O(\Delta x). \tag{3.2}$$

Since this occurs in (Δx) the largest term of the truncation error to the first power, we speak of a truncation error of first order $O(\Delta x)$ or a discretization of first order accuracy.

For the backward difference the same order of accuracy results from the Taylor series expansion

$$U_{i-1,j} = U_{i,j} - \Delta x \cdot \left(\frac{\partial U}{\partial x}\right)_{i,j} + \frac{(\Delta x)^2}{2} \cdot \left(\frac{\partial^2 U}{\partial x^2}\right)_{i,j} - \frac{(\Delta x)^3}{6} \cdot \left(\frac{\partial^3 U}{\partial x^3}\right)_{i,j} + \cdots \tag{3.3}$$

Resolved, the first derivative is

$$\left(\frac{\partial U}{\partial x}\right)_{i,j} = \underbrace{\frac{U_{i,j} - U_{i-1,j}}{\Delta x}}_{\text{Finite Difference}} \underbrace{+ \frac{(\Delta x)}{2} \cdot \left(\frac{\partial^2 U}{\partial x^2}\right)_{i,j} - \frac{(\Delta x)^2}{6} \cdot \left(\frac{\partial^3 U}{\partial x^3}\right)_{i,j} + \cdots}_{\text{Termination error}}$$

With the truncation error, the **first order backward difference is** obtained

$$\left(\frac{\partial U}{\partial x}\right)_{i,j} = \frac{U_{i,j} - U_{i-1,j}}{\Delta x} + O(\Delta x). \tag{3.4}$$

This means that these forward and backward differences are only first order accurate. This order of accuracy is not sufficient for continuous solutions.

In contrast, the central difference of second order is exact. If the difference of Eqs. 3.1 and 3.3 is formed, we get

$$U_{i+1,j} - U_{i-1,j} = 2 \cdot (\Delta x) \cdot \left(\frac{\partial U}{\partial x}\right)_{i,j} + 2 \cdot \frac{(\Delta x)^3}{6} \cdot \left(\frac{\partial^3 U}{\partial x^3}\right)_{i,j} + \cdots$$

Resolved according to the partial derivative we are looking for, we get the **central difference of the second order**

$$\left(\frac{\partial U}{\partial x}\right)_{i,j} = \frac{U_{i+1,j} - U_{i-1,j}}{2 \cdot \Delta x} + O(\Delta x)^2. \tag{3.5}$$

It is second order accurate because the largest truncation error term occurs $(\Delta x)^2$ with the forward and backward differences can also be formed with second order accuracy by including three points. For the **second order forward difference this** results in

$$\left(\frac{\partial U}{\partial x}\right)_{i,j} = \frac{-3 \cdot U_{i,j} + 4 \cdot U_{i+1,j} - U_{i+2,j}}{2 \cdot \Delta x} + O(\Delta x)^2 \tag{3.6}$$

and for the **second order backward difference** analogously

$$\left(\frac{\partial U}{\partial x}\right)_{i,j} = \frac{3 \cdot U_{i,j} - 4 \cdot U_{i-1,j} + U_{i-2,j}}{2 \cdot \Delta x} + O(\Delta x)^2. \tag{3.7}$$

In summary, for the discretization of the first derivative in the *x-direction*, the following possibilities for difference formation arise

$$\left(\frac{\partial U}{\partial x}\right)_{i,j} = \begin{cases} \dfrac{U_{i+1,j} - U_{i,j}}{\Delta x} + O(\Delta x) & \text{Forward 1st order} \\[2mm] \dfrac{-3 \cdot U_{i,j} + 4 \cdot U_{i+1,j} - U_{i+2,j}}{2 \cdot \Delta x} + O(\Delta x)^2 & \text{Forward 2nd order} \\[2mm] \dfrac{U_{i,j} - U_{i-1,j}}{\Delta x} + O(\Delta x) & \text{Backward 1st order} \\[2mm] \dfrac{3 \cdot U_{i,j} - 4 \cdot U_{i-1,j} + U_{i-2,j}}{2 \cdot \Delta x} + O(\Delta x)^2 & \text{Backward 2st order} \\[2mm] \dfrac{U_{i+1,j} - U_{i-1,j}}{2 \cdot \Delta x} + O(\Delta x)^2 & \text{Central 2nd order} \end{cases} \tag{3.8}$$

and correspondingly in *y-direction*

$$\left(\frac{\partial U}{\partial y}\right)_{i,j} = \begin{cases} \dfrac{U_{i,j+1} - U_{i,j}}{\Delta y} + O(\Delta y) & \text{Forward 1st order} \\[2mm] \dfrac{-3 \cdot U_{i,j} + 4 \cdot U_{i,j+1} - U_{i,j+2}}{2 \cdot \Delta y} + O(\Delta y)^2 & \text{Forward 2st order} \\[2mm] \dfrac{U_{i,j} - U_{i,j-1}}{\Delta y} + O(\Delta y) & \text{Backward 1st order} \\[2mm] \dfrac{3 \cdot U_{i,j} - 4 \cdot U_{i,j-1} + U_{i,j-2}}{2 \cdot \Delta y} + O(\Delta y)^2 & \text{Backward 2st order} \\[2mm] \dfrac{U_{i,j+1} - U_{i,j-1}}{2 \cdot \Delta y} + O(\Delta y)^2 & \text{Central 2nd order} \end{cases} \tag{3.9}$$

3.3.2 Discretization of the Second Derivatives

In the Navier-Stokes equations there are also second derivatives in the momentum equations and the energy equation. If, for example, the expressions for the stress terms from Eqs. 2.29–2.37 into the *x-momentum equation* Eq. 2.10, the following results

$$\frac{\partial}{\partial t}(\rho \cdot u) + \frac{\partial}{\partial x}(\rho \cdot u^2) + \frac{\partial}{\partial y}(\rho \cdot u \cdot v) + \frac{\partial}{\partial z}(\rho \cdot u \cdot w) + \frac{\partial p}{\partial x}$$
$$+ \frac{\partial}{\partial x}\left[\frac{2}{3}\mu \cdot \left(\frac{\partial u}{\partial x} + \frac{\partial v}{\partial y} + \frac{\partial w}{\partial z}\right) - 2 \cdot \mu \cdot \frac{\partial u}{\partial x}\right] - \frac{\partial}{\partial y}\left[\mu \cdot \left(\frac{\partial v}{\partial x} + \frac{\partial u}{\partial y}\right)\right]$$
$$- \frac{\partial}{\partial z}\left[\mu \cdot \left(\frac{\partial v}{\partial x} + \frac{\partial u}{\partial y}\right)\right] = \rho \cdot g_x.$$

If the friction terms are summarized at constant μ

$$\mu \cdot \left[\frac{2}{3} \cdot \left(\frac{\partial^2 u}{\partial x^2} + \frac{\partial^2 v}{\partial x \cdot \partial y} + \frac{\partial^2 w}{\partial x \cdot \partial z}\right) - 2 \cdot \frac{\partial^2 u}{\partial x^2} + \left(\frac{\partial^2 v}{\partial y \cdot \partial x} + \frac{\partial^2 u}{\partial y^2}\right) \quad - \left(\frac{\partial^2 v}{\partial z \cdot \partial x} + \frac{\partial^2 u}{\partial z \cdot \partial y}\right)\right],$$

so result the second derivatives $\frac{\partial^2 u}{\partial x^2}$, and so on. $\frac{\partial^2 v}{\partial x \cdot \partial y}$.

The finite differences for these second derivatives can again be derived from the Taylor series expansions of the forward and backward differences. If Eqs. 3.1 and 3.3 are added, we get

$$U_{i+1,j} + U_{i-1,j} = U_{i,j} + (\Delta x) \cdot \left(\frac{\partial U}{\partial x}\right)_{i,j} + \frac{(\Delta x)^2}{2} \cdot \left(\frac{\partial^2 U}{\partial x^2}\right)_{i,j} + \frac{(\Delta x)^3}{6} \cdot \left(\frac{\partial^3 U}{\partial x^3}\right)_{i,j} + \cdots$$

$$+ U_{i,j} - (\Delta x) \cdot \left(\frac{\partial U}{\partial x}\right)_{i,j} + \frac{(\Delta x)^2}{2} \left(\frac{\partial^2 U}{\partial x^2}\right)_{i,j} - \frac{(\Delta x)^3}{6} \cdot \left(\frac{\partial^3 U}{\partial x^3}\right)_{i,j} + \cdots$$

or

$$U_{i+1,j} + U_{i-1,j} = 2 \cdot U_{i,j} + (\Delta x)^2 \cdot \left(\frac{\partial^2 U}{\partial x^2}\right)_{i,j} + \frac{(\Delta x)^4}{12} \cdot \left(\frac{\partial^4 U}{\partial x^4}\right)_{i,j} + \cdots$$

Solved for the **second derivative in x-direction** results in

$$\left(\frac{\partial^2 U}{\partial x^2}\right)_{i,j} = \frac{U_{i+1,j} - 2 \cdot U_{i,j} + U_{i-1,j}}{(\Delta x)^2} + O(\Delta x)^2. \tag{3.10}$$

Analogous to this, the following results for the **second derivative in y-direction**

$$\left(\frac{\partial^2 U}{\partial y^2}\right)_{i,j} = \frac{U_{i,j+1} - 2 \cdot U_{i,j} + U_{i,j-1}}{(\Delta y)^2} + O(\Delta y)^2 \tag{3.11}$$

and for the **mixed second derivatives**

$$\left(\frac{\partial^2 U}{\partial x \cdot \partial y}\right)_{i,j} = \frac{U_{i+1,j+1} - U_{i+1,j-1} - U_{i-1,j+1} + U_{i-1,j-1}}{4 \cdot \Delta x \cdot \Delta y} + O\left[(\Delta x)^2, (\Delta y)^2\right]. \tag{3.12}$$

3.3.3 Notes on Spatial Discretization

- There are also difference expressions with higher order of accuracy. But then more net points are included.
- An accuracy order of two is usually sufficient and standard in commercial CFD programs. Therefore, central spatial differences accurate to second order are usually used for the first derivatives.
- However, the situation is different if discontinuities occur in the flow, such as compression shocks. Then, the central spatial differences lead to an error, since they assume a steady course of the flow variables. Modern CFD methods therefore automatically switch to one-sided differences of first or second order accuracy in shock environments. One speaks then of the upwind discretization. Upwind means that the discretization

Fig. 3.4 Mesh points at the solid edge

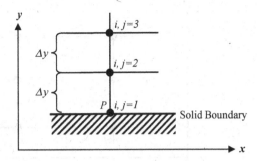

direction (forward or backward) is selected according to the physical propagation direction of disturbances. More on this in Sects. 3.5 and 5.4.

- At the edges of the computational domain such as at the inflow-, outflow and solid edges, only one-sided differences can be used because points outside the mesh domain are missing (Fig. 3.4). The forward difference according to Eq. 3.2 is here

$$\left(\frac{\partial U}{\partial y}\right)_1 = \frac{U_2 - U_1}{\Delta y} + O(\Delta y). \tag{3.13}$$

However, this difference is only of first order accuracy. To obtain second order accuracy, higher one-sided differences are used. The one-sided forward difference of second order is then analogous to Eq. 3.6, e.g.

$$\left(\frac{\partial U}{\partial y}\right)_1 = \frac{-3 \cdot U_1 + 4 \cdot U_2 - U_3}{2 \cdot \Delta y} + O(\Delta y)^2, \tag{3.14}$$

where now three netpoints are used instead of two.

3.4 Time Discretization

The time discretization of the partial derivative is $\partial/\partial t$ similar to the spatial discretization. Figure 3.5 shows three different time levels $n - 1, n, n + 1$, where the solution at time $n + 1$ is sought and all previous solutions at time $n, n - 1, n - 2$ etc. are known.

For the temporal discretization, analogous to the spatial differences, a **temporal forward difference with first order** precision

$$\left(\frac{\partial U}{\partial t}\right)_i^n = \frac{U_i^{n+1} - U_i^n}{\Delta t} + O(\Delta t) \tag{3.15}$$

or a **time central difference of second order** accuracy can be chosen

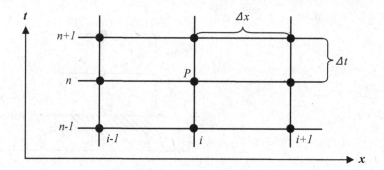

Fig. 3.5 The time levels n for the temporal discretization

$$\left(\frac{\partial U}{\partial t}\right)_i^n = \frac{U_i^{n+1} - U_i^{n-1}}{2 \cdot \Delta t} + O(\Delta t)^2. \tag{3.16}$$

A backward difference is not possible, since the solution at time $n + 1$ is sought.

Which temporal discretization is used depends on whether temporal accuracy is important or not. A distinction is made between:

- of a time asymptotic or stationary solution (Eq. 3.15)
- and a time-accurate or transient solution (Eq. 3.16).

3.4.1 Time Asymptotic or Stationary Solutions

The five conservation equations for mass, momentum and energy also contain the time derivatives. For stationary processes, they could actually be neglected, so that only the stationary conservation equations with the spatial derivatives remain. However, this has the disadvantage that different solution methods would have to be used for subsonic and supersonic regions, since the differential equations change their type (elliptic, hyperbolic) depending on whether a subsonic or supersonic flow exists.

Therefore, it has proven expedient to solve the transient conservation equations also for stationary problems. They then always remain hyperbolic and can be solved for both subsonic and supersonic regions with a single solution procedure. The time derivative is then no longer used to resolve the physics at each time point, but is only used to numerically solve the steady-state conservation equations. Therefore, an accuracy of first order is sufficient and Eq. 3.15 can be used.

Starting from a relatively imprecise initial solution at time $t = 0$ or at time level $n = 0$, a solution is calculated again and again at the next time level $n + 1$ until the solutions no longer differ at two successive time steps. This is called a converged solution. If this is the case

$$U_i^{n+1} \cong U_i^n$$

and the time derivative is

$$\left(\frac{\partial U}{\partial t}\right)_i^n = \frac{U_i^{n+1} - U_i^n}{\Delta t} \cong 0,$$

i.e. the steady-state conservation equations are satisfied.

Since only the solution in the converged state is of interest and the intermediate solutions are not of interest, the time step can be chosen so large that the solution procedure converges as fast as possible. The time step has (almost) nothing to do with physics, but is a purely numerical value ($\Delta t_{num.}$). Modern CFD programs calculate it automatically, mostly even for each mesh cell locally, in order to achieve an optimal convergence behavior.

3.4.2 Time-Accurate or Transient Solutions

The situation is different if transient effects have to be resolved. Since the temporal accuracy (like the spatial accuracy) should be of at least second order, Eq. 3.16 must be used. The time step must then be chosen so that it corresponds to the physical problem ($\Delta t_{phys.}$), i.e. still captures the temporal changes of interest. The physical time step must be specified by the user.

The computation times and the memory requirements are often very large for transient solutions. On the one hand, the time steps must usually be small in order to be able to resolve the temporal changes and, on the other hand, the solution must be stored at every point in time.

Most commercial programs can calculate both steady-state and transient processes, it is only necessary to change one input parameter. The program then selects the correct forward difference.

3.5 Difference Equations

3.5.1 Derivation

Difference equations result from the differential equations, if for the spatial and temporal derivatives the difference expressions are used. For example, from the Navier-Stokes or Euler equation in vector form (Eq. 2.18), where the vector arrows are now omitted for the sake of simplicity, results

$$\frac{\partial}{\partial t}U + \frac{\partial}{\partial x}E + \frac{\partial}{\partial y}F + \frac{\partial}{\partial z}G = Q$$

with central spatial differences in the three spatial directions analogous to Eq. 3.5

$$\left(\frac{\partial}{\partial x}E\right)_{i,j,k} = \frac{E_{i+1,j,k} - E_{i-1,j,k}}{2 \cdot \Delta x} + O(\Delta x)^2, \tag{3.17}$$

$$\left(\frac{\partial}{\partial y}F\right)_{i,j,k} = \frac{F_{i,j+1,k} - F_{i,j-1,k}}{2 \cdot \Delta y} + O(\Delta y)^2, \tag{3.18}$$

$$\left(\frac{\partial}{\partial z}G\right)_{i,j,k} = \frac{G_{i,j,k+1} - G_{i,j,k-1}}{2 \cdot \Delta z} + O(\Delta z)^2 \tag{3.19}$$

- and the one-sided temporal forward difference from Eq. 3.15

$$\left(\frac{\partial}{\partial t}U\right)_{i,j,k}^n = \frac{U_{i,j,k}^{n+1} - U_{i,j,k}^n}{\Delta t} + O(\Delta t)$$

following difference equation

$$\frac{U_{i,j,k}^{n+1} - U_{i,j,k}^n}{\Delta t} + O(\Delta t) + \frac{E_{i+1,j,k}^n - E_{i-1,j,k}^n}{2 \cdot \Delta x} + O(\Delta x)^2$$
$$+ \frac{F_{i,j+1,k}^n - F_{i,j-1,k}^n}{2 \cdot \Delta y} + O(\Delta y)^2 + \frac{G_{i,j,k+1}^n - G_{i,j,k-1}^n}{2 \cdot \Delta z} + O(\Delta z)^2 = Q_{i,j,k}^n. \tag{3.20}$$

As can be $O(\Delta t)$ seen $O(\Delta x)^2$, $O(\Delta y)^2$, $O(\Delta z)^2$ from the truncation errors and, this equation is first order exact in time and second order exact in space. The truncation error terms will be omitted from now on, since they only serve to show the order of accuracy.

If Eq. 3.20 is solved for the unknown quantity at time $n + 1$, we obtain the **difference equation**

$$U_{i,j,k}^{n+1} = U_{i,j,k}^n - \Delta t$$
$$\cdot \left[\frac{E_{i+1,j,k}^n - E_{i-1,j,k}^n}{2 \cdot \Delta x} + \frac{F_{i,j+1,k}^n - F_{i,j-1,k}^n}{2 \cdot \Delta y} + \frac{G_{i,j,k+1}^n - G_{i,j,k-1}^n}{2 \cdot \Delta z} - Q_{i,j,k}^n\right]. \tag{3.21}$$

The conservation variable U at grid point i, j, k *at* time $n + 1$ can be calculated in principle with Eq. 3.21, since the flux quantities E, F, G and the source term Q at time n *are* known.

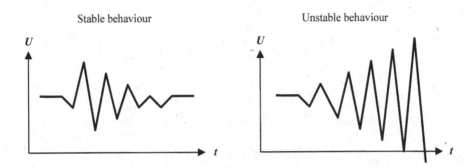

Fig. 3.6 Course of the flow variable u over time for a stable and an unstable solution

Unfortunately, Eq. 3.21 does not lead to a stable solution, since the truncation error increases from time step to time step, or the fluctuations of the flow magnitude U increase, as Fig. 3.6 shows.

Therefore, the stability of the difference equation Eq. 3.21 must be improved. Now what is a stable solution and how do I achieve it? To be able to answer this question, the following three terms must be explained first.

3.5.2 Consistency, Stability and Convergence

Consistency

The difference equations are consistent if they merge Δt, Δx, Δy, $\Delta z \to 0$ into the differential equations for or if their truncation errors $O(\Delta t)$, $O(\Delta x)$, $O(\Delta x)^2$, $O(\Delta y)^2$, $O(\Delta z)^2$ etc. for Δt, Δx, Δy, $\Delta z \to 0$ become zero. This ensures that the difference equations also represent the physical differential equations and that no terms are missing or too many.

Stability

A numerical solution method is stable if the truncation errors, which are neglected in the numerical solution, become smaller and smaller. The numerical solution then satisfies the differential equations. In mathematics there are numerous methods for the investigation of the stability behavior of differential equations. The best known is the Von Neumann stability analysis, which investigates whether small perturbations are damped or fanned.

Convergence

A numerical solution is convergent if it satisfies the differential equations. This is checked, for example, in time-stepping methods by means of the so-called residual. The residual is a numerical value for each conservation equation, which indicates to what extent the stationary conservation equation is satisfied. If it has decreased by four to five orders of magnitude, then in practice one speaks of a convergent solution (theoretically, the residual should be exactly zero).

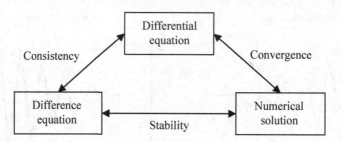

Fig. 3.7 Relationships between consistency, stability and convergence

In summary, it can be said that for a physical solution it is at least necessary that

- the discretization is consistent,
- the solution procedure stable
- and the solution is convergent.

Figure 3.7 shows these relationships again graphically.

How can we now modify the difference equation Eq. 3.21 so that the solution procedure becomes stable? Three possibilities are used in practice:

1. By introducing an additive numerical viscosity, the truncation errors are damped and the solution procedure becomes stable.
2. Instead of the central spatial differences, one-sided spatial differences are used (upwind method).
3. The spatial derivatives are not formed at time n (explicit method), but at time $n + 1$ (implicit method).

3.5.3 Additive Numerical Viscosity

The introduction of additive numerical viscosity is best shown by the one-dimensional difference equation with no source term, which follows from Eq. 3.21:

$$U_i^{n+1} = U_i^n - \frac{1}{2} \cdot \frac{\Delta t}{\Delta x} \cdot \left[E_{i+1}^n - E_{i-1}^n \right]. \tag{3.22}$$

As already mentioned above, this difference equation with a central spatial difference is not stable, but an additional term must be added, the so-called numerical viscosity. It is also called artificial viscosity or numerical damping or numerical dissipation in contrast to the physical viscosity in the friction terms of the Navier-Stokes equations.

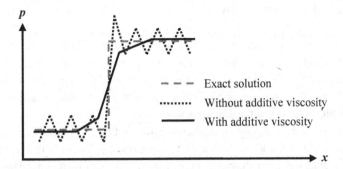

Fig. 3.8 Principle pressure curve at a compression joint (exact, without and with additive numerical viscosity)

Figure 3.8 shows the principal course of the static pressure at a compression joint with and without additive numerical viscosity in comparison to the exact solution. Without numerical viscosity, the solution oscillates, since the process becomes unstable at the joint. These oscillations are suppressed by the numerical viscosity.

The **additive numerical viscosity** is a term that corresponds to a second and fourth derivative and looks similar to the physical viscosity

$$U_i^{n+1} = U_i^n - \frac{1}{2} \cdot \frac{\Delta t}{\Delta x} \cdot \left[E_{i+1}^n - E_{i-1}^n\right] - \underbrace{\varepsilon_2 \cdot \left[U_{i+1}^n - 2 \cdot U_i^n + U_{i-1}^n\right]}_{\text{Numerical viscosity 2nd order}}$$
$$- \underbrace{\varepsilon_4 \cdot \left[U_{i+2}^n - 4 \cdot U_{i+1}^n + 6 \cdot U_i^n - 4 \cdot U_{i-1}^n + U_{i-2}^n\right]}_{\text{Numerical viscosity 4th order}}. \tag{3.23}$$

The prefactors and ε_2 are ε_4 to be predefined. They must be small enough not to distort the solution noticeably, but large enough to make the procedure stable, i.e. to damp it sufficiently. Typical values are and $\varepsilon_2 \approx 1/4 \varepsilon_4 \approx 1/256$.

While the numerical viscosity of the fourth order dampens the disturbances in the entire computational domain and thus makes the difference equation stable, the numerical viscosity of the second order acts rather in the area of discontinuities such as compression shocks. It ensures that the oscillations before and after the shock are significantly reduced and the shock is more sharply resolved or less smeared. Therefore, the fourth order term is usually switched off at the joint, while the second order term only acts at the joint. This is controlled by a term that depends on the pressure gradient and is large at the joint, but small in the remaining computational domain.

3.5.4 Upwind Discretization

Another way to make the difference equation stable is the so-called upwind discretization. In this case, one-sided forward and backward differences are used instead of the central spatial discretization. This eliminates the need to add the numerical viscosity. It can be shown mathematically that the upwind discretization can be converted into a central discretization plus additive numerical viscosity.

The special feature of the upwind discretization is that the propagation direction of physical disturbances such as sound waves or compression shocks is taken into account. Depending on whether the disturbance comes from upstream or downstream, one-sided forward or backward differences are used. The discretization is thus always in the direction of flow, hence the name "upwind".

If the one-sided spatial differences are used, the following first **order upwind difference equations are obtained**.

$$U_i^{n+1} = U_i^n - \frac{\Delta t}{\Delta x} \cdot \left[E_{i+1}^n - E_i^n \right], \tag{3.24}$$

$$U_i^{n+1} = U_i^n - \frac{\Delta t}{\Delta x} \cdot \left[E_i^n - E_{i-1}^n \right]. \tag{3.25}$$

Equation 3.24 uses the first order forward difference, while Eq. 3.25 uses the first order backward difference. In order to decide which of the two equations must be used, the direction of propagation of the flow is first calculated, e.g. using characteristics theory. Therefore, the computational effort is higher than for the central spatial difference.

Unfortunately, first order accuracy is usually not accurate enough, so **second order** one-way **upwind difference equations** are used for more accurate methods

$$U_i^{n+1} = U_i^n - \frac{\Delta t}{2 \cdot \Delta x} \cdot \left[- 3 \cdot E_i^n + 4 \cdot E_{i+1}^n - E_{i+2}^n \right], \tag{3.26}$$

$$U_i^{n+1} = U_i^n - \frac{\Delta t}{2 \cdot \Delta x} \cdot \left[3 \cdot E_i^n - 4 \cdot E_{i-1}^n + E_{i-2}^n \right]. \tag{3.27}$$

Equation 3.26 is again a forward difference, but now second order, while Eq. 3.27 is the backward difference second order.

These methods are also called high-resolution methods, since they are upwind methods of second order accuracy. The impact resolution is not as good as with the first order upwind methods, but the accuracy in the remaining flow field is significantly better, more on this in Sect. 5.5.

3.5.5 Explicit and Implicit Discretization

The calculation of the flow terms can be done explicitly or implicitly, depending on the time level on which they are formed. The difference can be seen in the one-dimensional difference equation (Eq. 3.22). In the explicit difference scheme, the flow terms are formed nE at the time point

$$U_i^{n+1} = U_i^n - \frac{1}{2} \cdot \frac{\Delta t}{\Delta x} \cdot \left[E_{i+1}^n - E_{i-1}^n \right].$$

Therefore, the sought quantity U_i^{n+1} at time $n + 1$ can be calculated directly, i.e. explicitly from the known values at time n. This so-called explicit difference scheme is shown schematically in Fig. 3.9.

In contrast to this, in the so-called implicit difference scheme, the flow terms E are formed at the still unknown time $n + 1$

$$U_i^{n+1} + \underbrace{\frac{1}{2} \cdot \frac{\Delta t}{\Delta x}}_{c} \cdot \left[E_{i+1}^{n+1} - E_{i-1}^{n+1} \right] = U_i^n. \tag{3.28}$$

The conservation variable of interest U_i^{n+1} can now no longer be calculated directly at time $n + 1$, since the flux variables and E_{i+1}^{n+1} are still unknown E_{i-1}^{n+1} at time $n + 1$. The implicit difference scheme is shown schematically in Fig. 3.10.

U_i^{n+1} at network point i at time $n + 1$ can now only be calculated if such equations are set up for all network points and the resulting system of equations is solved

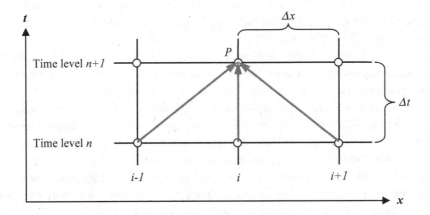

Fig. 3.9 An explicit temporal difference scheme

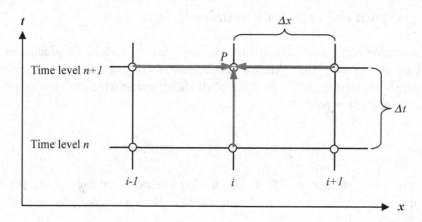

Fig. 3.10 An implicit temporal difference scheme

$$\vdots$$
$$c \cdot E_{i-1}^{n+1} + U_i^{n+1} - c \cdot E_{i+1}^{n+1} = U_i^n$$
$$c \cdot E_i^{n+1} + U_{i+1}^{n+1} - c \cdot E_{i+2}^{n+1} = U_{i+1}^n \qquad (3.29)$$
$$c \cdot E_{i+1}^{n+1} + U_{i+2}^{n+1} - c \cdot E_{i+3}^{n+1} = U_{i+2}^n$$
$$\vdots$$

This leads with central spatial discretization to a so-called tridiagonal system of equations, since only the main diagonal and the two left and right adjacent diagonals of the matrix are *M* occupied and all other values are zero.

In order to solve this system of equations, the terms E_i^{n+1}/U_i^{n+1} in the matrix *M* must be transformed in such a way that they can be calculated from values that are already known. This is achieved by the so-called linearization, the mathematical formulation of which can be found in the technical literature [3].

On the other hand, the matrix *M* must be inverted or the system of equations must be solved iteratively using a relaxation method. Both are relatively time-consuming and the computation time per time step Δt is greater than with explicit methods.

On the other hand, implicit methods need significantly fewer iterations to reach the convergent solution, since they are more stable and can use significantly larger time steps. As a result, the total computing times of implicit methods are often lower than those of explicit methods, despite the higher computing time per iteration.

However, in modern CFD methods, explicit discretization is mostly used because it is simpler and more flexible, and numerous methods for convergence acceleration are now available, such as local time-stepping and multigrid techniques.

It should also be mentioned that although the implicit discretization is stable for linear problems, it can become unstable for nonlinear problems such as flows with compressional shocks. Therefore, an additive numerical viscosity or the upwind discretization is also used for implicit solution methods in order to be able to achieve a good stability and a good shock resolution even for shocked flows.

3.5.6 CFL Number

Equation 3.28 contains the ratio of time step to computational mesh spacing $\Delta t \Delta x$. A stability investigation [3] shows that this ratio $\Delta t / \Delta x$ must not be chosen arbitrarily, since otherwise the termination errors become too large and the method becomes unstable.

Physically, this can be explained by the propagation direction of disturbances, the so-called characteristics. The propagation speed of information in the computing network must always be smaller than the physical propagation speed of a disturbance such as a sound wave. Therefore, the speed of sound is also included. This results in the so-called **Courant-Friedrichs-Levy number (CFL)**, which is defined as follows:

$$\text{CFL} = a \cdot \frac{\Delta t}{\Delta x}. \tag{3.30}$$

The time step is coupled Δt to the computational mesh size Δx and the sound velocity a. *For* small mesh sizes, the time step or the CFL number must therefore be smaller than for large mesh sizes.

A stability investigation shows that for purely explicit methods, the CFL number must always be less than 1 in order to achieve a stable and convergent solution. In contrast, implicit methods are stable even with much larger CFL numbers (e.g. 10–1000 depending on the solution method and complexity of the flow problem).

3.5.7 Summary

Finally to Chap. 3 a summary of the most important discretization possibilities and their advantages and disadvantages is given in Tables 3.1 and 3.2.

Table 3.1 Most important properties of the spatial discretization

Spatial discretization	
Upwind 1st order accurate	One-sided differences 1st order Poor accuracy, as only 1st order Shocks are smeared out Very stable
Central 2nd order accurate	Central differences 2nd order Good accuracy, as 2nd order Oscillations at discontinuities as shocks Stable only with added numerical viscosity
High resolution 2nd order accurate	One-sided differences 2nd order Good accuracy, as 2nd order Good resolution of discontinuities as shocks Stable

Table 3.2 Most important properties of the temporal discretization

Time Discretisation	
1st order accurate	Für stationäre Probleme ausreichend
2nd order accurate	Für instationäre Probleme notwendig
Explicit method	Flussterme werden zum Zeitpunkt n gebildet Gesuchte Größe kann explizit ausgerechnet werden Zeitschritt pro Iteration muss aus Stabilitätsgründen klein sein (CFL<1), deshalb sind viele Zeitschritte notwendig bis zur konvergenten Lösung
Implicit method	Flussterme werden zum Zeitpunkt n+1 berechnet Gesuchte Größe kann nur über ein Gleichungssystem (implizit) berechnet werden, dessen Matrix invertiert werden muss (höhere Rechenzeit pro Iteration) Zeitschritt pro Iteration ist aber größer (CFL>>1), deshalb weniger Zeitschritte notwendig bis zur konvergenten Lösung

Computational Meshes

4

4.1 Aim of This Chapter

The discretized conservation equations can only be solved at defined grid points. For this, a mesh is necessary, which fills the area in which the flow is to be calculated. The convergence behaviour of the calculation and the accuracy of the solution depend decisively on the quality of the mesh. As fine as necessary, as coarse as possible, is the motto here. On the one hand, important flow details should be captured exactly, on the other hand, the computation times should be acceptable. Often, for more complex applications, the number of mesh points is chosen just large enough so that the calculation run sent in the afternoon can be evaluated the next morning.

Although modern mesh generation programs are becoming more and more user-friendly, the user must still have some experience to set the global and local compaction parameters to meet the accuracy and computation time requirements. In this chapter the basics of mesh generation are presented.

You should then be able to answer the following questions:

1. Why do you need a mesh?
2. What requirements should a good mesh meet?
3. What is the advantage of curvilinear meshes?
4. What is meant by an O, C, H mesh? What are their advantages and disadvantages?
5. Why are meshes refined at the solid boundary? What happens in frictionless computation?
6. With how many net points should the boundary layer be resolved at least?
7. What is the advantage of block structured meshs?
8. What is meant by adaptive meshs? What advantage do they have?

© The Author(s), under exclusive license to Springer Fachmedien Wiesbaden GmbH, part of Springer Nature 2022
S. Lecheler, *Computational Fluid Dynamics*,
https://doi.org/10.1007/978-3-658-38453-1_4

9. What is the advantage of unstructured meshs?

10. How should an ideal mesh cell be, more right-angled or more skew-angled?

4.2   Overview

The discretized differential equations result in the so-called difference equation, which was derived in Chap. 3 (Eq. 3.21):

$$U_{i,j,k}^{n+1} = U_{i,j,k}^{n} - \Delta t \cdot \left[\frac{E_{i+1,j,k}^{n} - E_{i-1,j,k}^{n}}{2 \cdot \Delta x} + \frac{F_{i,j+1,k}^{n} - F_{i,j-1,k}^{n}}{2 \cdot \Delta y} + \frac{G_{i,j,k+1}^{n} - G_{i,j,k-1}^{n}}{2 \cdot \Delta z} - Q_{i,j,k}^{n} \right].$$

In order to solve it, the conservation vector U, the flux terms E, F, G and the source term Q at the interpolation points (i, j, k), $(i + 1, j, k)$, etc.$(i - 1, j, k)$ must be known. A mesh is necessary for this. It fills the area in which the flow is to be calculated.

Figure 4.1 shows such a mesh schematically with the geometry and some boundaries.

The geometry, such as the vehicle or the aircraft wing, is usually imported from a CAD program. Modern commercial CFD programs have interfaces to all common CAD programs. For simpler geometries, it can also be specified as a set of points or as an analytical function. For example, airfoil profiles are often specified as an analytical shape, since they can then be modified and optimized on the basis of a few parameters. Practical tips for this are given in Chap. 6.

After the geometry data has been read in, the boundaries of the calculation area must be defined. Often symmetry planes can be exploited to save mesh points and computation

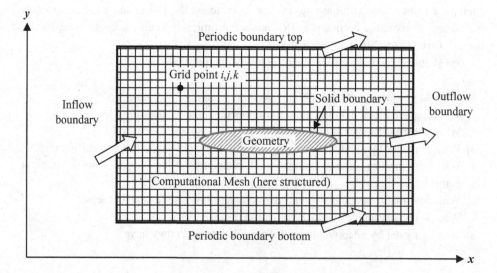

Fig. 4.1 Definitions of the mesh

time. The so-called solid boundary is the surface of the geometry that is impermeable to the flow. Furthermore, the inflow and outflow boundaries and the periodic boundaries have to be defined. At them the flow enters or leaves the computational domain.

Only now can the actual generation of the mesh take place. There are different mesh forms, which are described in more detail in this chapter, including their advantages and disadvantages.

The final step is the mesh adaptation, where the mesh is adapted to the gradients of the flow (e.g. boundary layer, compaction shocks) and the geometry (e.g. strong curvatures at kinks and edges). Figure 4.2 gives an overview of the mesh generation process.

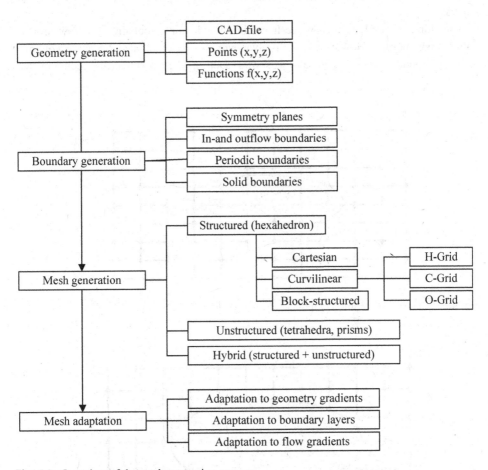

Fig. 4.2 Overview of the mesh generation process

4.3   Structured Meshes

4.3.1   Cartesian Meshe

Cartesian meshes are very easy to generate and have rectangular cells that result in a small truncation error and thus good accuracy. Figure 4.3 shows a geometry with a rectangular Cartesian mesh. What stands out? Some mesh points lie within the geometry in the dotted area, where there is no flow at all. No flow quantities can be calculated here. On the surface of the geometry there are only a few or no mesh points, which are also distributed rather randomly. This is bad, since it is the flow quantities on the surface of the profile that are of interest.

Therefore, a purely Cartesian mesh is not suitable for calculating the flow around arbitrary geometries. However, this disadvantage can be eliminated by using contour-matched boundary cells. Figure 4.4 shows such a mesh. While the cells in the

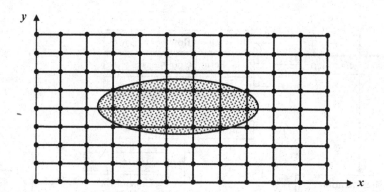

Fig. 4.3 Schematic of a Cartesian mesh

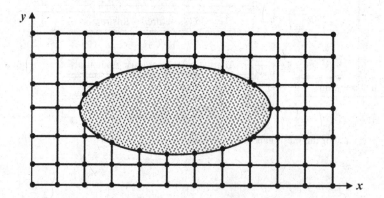

Fig. 4.4 Schematic of a Cartesian mesh with contour-matched boundary cells

computational domain are purely rectangular, the cells close to the wall are adapted to the contour. However, this results in irregular cells with three, four or five corners. This must be taken into account in the solution algorithm and in the boundary conditions.

However, most numerical flow calculation methods use curvilinear meshes. They can be adapted much better to more complex geometries, as will be shown in Chap. 5.

4.3.2 Curvilinear Meshes

With curvilinear meshes, the mesh lines adapt to the wall contour. The coordinate system is now no longer Cartesian x, y, z, but curvilinear ξ, η, ζ. Depending on the shape of the mesh lines, a distinction is made between O, C and H meshes.

In the **O-grid** shown in Fig. 4.5, the grid lines run around the $\xi = $ const. geometry. The mesh lines $\eta = $ const. run away from the geometry in a star shape and are approximately perpendicular to the mesh lines $\xi = $ const. An advantage of the O-grids is that the boundary layer at the geometry itself can be resolved very well. However, for transversely flowed airfoils, the resolution of the flow wake downstream of the trailing edge is not as good, resulting in skewed mesh cells that lead to higher truncation error and reduced accuracy. O-frame meshes are more suitable for thick round trailing edges as in turbine airfoils.

Figure 4.6 shows a contour-fitted **curvilinear C-mesh.** The mesh lines run around the geometry $\xi = $ const. from on side of the downstream boundary to the other side of the downstream boundary. The mesh lines run away $\eta = $ const from the geometry. With the C mesh, both the profile boundary layer and the profile wake downstream of the trailing edge can be well resolved. It is therefore more suitable for pointed trailing edges such as compressor profiles.

In the **H-grid** (Fig. 4.7), the grid lines run $\xi = $ const. from the inflow boundary along the profile to the outflow boundary. The other mesh lines $\eta = $ const. run from the lower

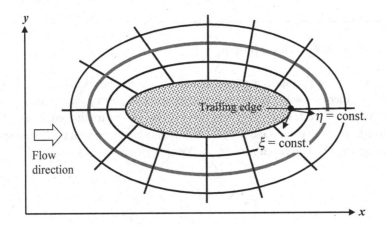

Fig. 4.5 Schematic of a curvilinear O-mesh

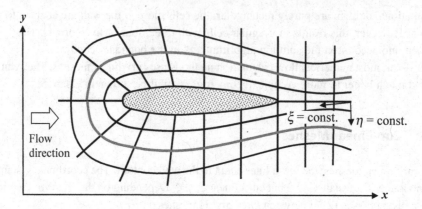

Fig. 4.6 Schematic of a curvilinear C-mesh

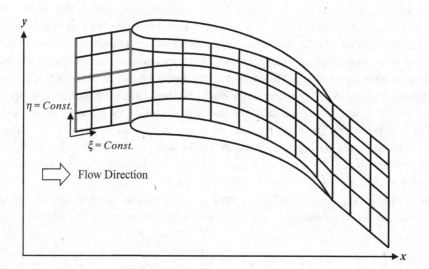

Fig. 4.7 Schematic of a curvilinear H-mesh

periodic boundary to the upper periodic boundary. At the profile, the mesh is usually compacted in order to resolve the boundary layer well. These compacted mesh lines also extend upstream of the leading edge and downstream of the trailing edge. The latter is desired in order to resolve the wake well. In contrast, the compaction in the inflow upstream of the leading edge is unnecessary, since the flow there is very uniform.

4.3.3 The Transformation of Coordinates into Curvilinear Coordinates

On these curvilinear meshes, the Cartesian difference equation (Eq. 3.21) can no longer be solved because the mesh is no longer Cartesian. Therefore a coordinate transformation must be carried out

$$
\begin{aligned}
\xi &= \xi(x, y, z, t), \\
\eta &= \eta(x, y, z, t), \\
\zeta &= \zeta(x, y, z, t), \\
\tau &= \tau(t).
\end{aligned}
\tag{4.1}
$$

The three new spatial coordinates ξ, η, ζ are functions of the three Cartesian spatial coordinates x, y, z. In the case of moving meshs they are also a function of time τ, where holds $\tau = t$. This coordinate transformation is carried out analytically with the differential equation from Chap. 2, since then no errors arise. This can be shown relatively simply at the example of the two-dimensional stationary transformation

$$
\begin{aligned}
\xi &= \xi(x, y), \\
\eta &= \eta(x, y).
\end{aligned}
\tag{4.2}
$$

For the partial derivatives $\partial/\partial x$, the following expressions are then obtained with the derivative with $\partial/\partial y$ respect to each of the two variables ξ, η

$$
\begin{aligned}
\frac{\partial}{\partial x} &= \left(\frac{\partial}{\partial \xi}\right) \cdot \left(\frac{\partial \xi}{\partial x}\right) + \left(\frac{\partial}{\partial \eta}\right) \cdot \left(\frac{\partial \eta}{\partial x}\right), \\
\frac{\partial}{\partial y} &= \left(\frac{\partial}{\partial \xi}\right) \cdot \left(\frac{\partial \xi}{\partial y}\right) + \left(\frac{\partial}{\partial \eta}\right) \cdot \left(\frac{\partial \eta}{\partial y}\right).
\end{aligned}
\tag{4.3}
$$

These partial derivatives for $\partial/\partial x$, according to Eq. 4.3 $\partial/\partial y$ could now simply be substituted into the differential equations, resulting $\partial/\partial \eta$ in differential equations in curvilinear coordinates with the partial derivatives $\partial/\partial \xi$. Unfortunately, the so-called metric derivatives $\partial\xi/\partial x$, $\partial\eta/\partial x$, $\partial\xi/\partial y$, cannot be $\partial\eta/\partial y$ computed because the curvilinear coordinates are usually ξ, η not known (they are only known if there are analytic relations between and ξ, η x, y).

On the other hand, the reversal lines $\partial x/\partial \xi$, $\partial y/\partial \xi$, can $\partial x/\partial \eta$ be $\partial y/\partial \eta$ calculated, since the Cartesian coordinates are known

$$\frac{\partial}{\partial x} = \frac{1}{J} \cdot \left[\left(\frac{\partial}{\partial \xi}\right) \cdot \left(\frac{\partial y}{\partial \eta}\right) - \left(\frac{\partial}{\partial \eta}\right) \cdot \left(\frac{\partial y}{\partial \xi}\right) \right],$$

$$\frac{\partial}{\partial y} = \frac{1}{J} \cdot \left[- \left(\frac{\partial}{\partial \xi}\right) \cdot \left(\frac{\partial x}{\partial \eta}\right) + \left(\frac{\partial}{\partial \eta}\right) \cdot \left(\frac{\partial x}{\partial \xi}\right) \right],$$

(4.4)

with the so-called Jacoby determinant

$$J = \text{Det} \begin{bmatrix} \dfrac{\partial x}{\partial \xi} & \dfrac{\partial x}{\partial \eta} \\[8pt] \dfrac{\partial y}{\partial \xi} & \dfrac{\partial y}{\partial \eta} \end{bmatrix} = \left(\frac{\partial x}{\partial \xi}\right) \cdot \left(\frac{\partial y}{\partial \eta}\right) - \left(\frac{\partial x}{\partial \eta}\right) \cdot \left(\frac{\partial y}{\partial \xi}\right).$$

(4.5)

For those interested, here is a short proof that Eqs. 4.4 and 4.3 are the same. Less interested people can go directly to Eq. 4.15.

For the transformation and $\xi = \xi(x, y)$ the total differentials are $\eta = \eta(x, y)$

$$d\xi = \left(\frac{\partial \xi}{\partial x}\right) \cdot dx + \left(\frac{\partial \xi}{\partial y}\right) \cdot dy$$

$$d\eta = \left(\frac{\partial \eta}{\partial x}\right) \cdot dx + \left(\frac{\partial \eta}{\partial y}\right) \cdot dy$$

or

$$\begin{bmatrix} d\xi \\ d\eta \end{bmatrix} = \begin{bmatrix} \dfrac{\partial \xi}{\partial x} & \dfrac{\partial \xi}{\partial y} \\[8pt] \dfrac{\partial \eta}{\partial x} & \dfrac{\partial \eta}{\partial y} \end{bmatrix} \cdot \begin{bmatrix} dx \\ dy \end{bmatrix}.$$

(4.6)

Analogously, the total differentials are $x = x(\xi, \eta)$ formed for $y = y(\xi, \eta)$ the inverse transformation and

$$dx = \left(\frac{\partial x}{\partial \xi}\right) \cdot d\xi + \left(\frac{\partial x}{\partial \eta}\right) \cdot d\eta$$

$$dy = \left(\frac{\partial y}{\partial \xi}\right) \cdot d\xi + \left(\frac{\partial y}{\partial \eta}\right) \cdot d\eta$$

or

$$\begin{bmatrix} dx \\ dy \end{bmatrix} = \begin{bmatrix} \dfrac{\partial x}{\partial \xi} & \dfrac{\partial x}{\partial \eta} \\[8pt] \dfrac{\partial y}{\partial \xi} & \dfrac{\partial y}{\partial \eta} \end{bmatrix} \cdot \begin{bmatrix} d\xi \\ d\eta \end{bmatrix}.$$

(4.7)

If now Eq. 4.7 is inverted or solved $d\eta$ to and, $d\xi$ then with the rules of matrix inversion we get

$$\begin{bmatrix} d\xi \\ d\eta \end{bmatrix} = \begin{bmatrix} \dfrac{\partial x}{\partial \xi} & \dfrac{\partial x}{\partial \eta} \\[8pt] \dfrac{\partial y}{\partial \xi} & \dfrac{\partial y}{\partial \eta} \end{bmatrix}^{-1} \cdot \begin{bmatrix} dx \\ dy \end{bmatrix} = \frac{1}{J} \cdot \begin{bmatrix} \dfrac{\partial y}{\partial \eta} & -\dfrac{\partial x}{\partial \eta} \\[8pt] -\dfrac{\partial y}{\partial \xi} & \dfrac{\partial x}{\partial \xi} \end{bmatrix} \cdot \begin{bmatrix} dx \\ dy \end{bmatrix},$$

(4.8)

with the so-called Jacoby determinant J according to Eq. 4.5.

A comparison of the individual terms of Eqs. 4.8 and 4.6 gives

$$\frac{\partial \xi}{\partial x} = \frac{1}{J} \cdot \frac{\partial y}{\partial \eta}, \tag{4.9}$$

$$\frac{\partial \eta}{\partial x} = -\frac{1}{J} \cdot \frac{\partial y}{\partial \xi}, \tag{4.10}$$

$$\frac{\partial \xi}{\partial y} = -\frac{1}{J} \cdot \frac{\partial x}{\partial \eta}, \tag{4.11}$$

$$\frac{\partial \eta}{\partial y} = \frac{1}{J} \cdot \frac{\partial x}{\partial \xi}, \tag{4.12}$$

which would prove that Eqs. 4.4 and 4.3 are identical.

Most numerical flow computation methods use this transformation from Cartesian to curvilinear coordinates, because curvilinear meshes are used for arbitrary geometries. In the following, this transformation is shown by the example of the conservation equations in vector form. They are in Cartesian coordinates according to Eq. 2.18

$$\frac{\partial}{\partial t} \vec{U} + \frac{\partial}{\partial x} \vec{E} + \frac{\partial}{\partial y} \vec{F} + \frac{\partial}{\partial z} \vec{G} = \vec{Q}.$$

The principle of the transformation can be shown more simply if the third dimension, the source term and the vector symbol are dispensed with

$$\frac{\partial}{\partial t} U + \frac{\partial}{\partial x} E + \frac{\partial}{\partial y} F = 0. \tag{4.13}$$

If the partial derivatives in Eq. 4.13 are replaced by the metric relations given in Eq. 4.4, the result is given by $t = \tau$

$$\left(\frac{\partial U}{\partial \tau}\right) + \frac{1}{J} \cdot \left[\left(\frac{\partial E}{\partial \xi}\right) \cdot \left(\frac{\partial y}{\partial \eta}\right) - \left(\frac{\partial E}{\partial \eta}\right) \cdot \left(\frac{\partial y}{\partial \xi}\right)\right]$$
$$+ \frac{1}{J} \cdot \left[-\left(\frac{\partial F}{\partial \xi}\right) \cdot \left(\frac{\partial x}{\partial \eta}\right) + \left(\frac{\partial F}{\partial \eta}\right) \cdot \left(\frac{\partial x}{\partial \xi}\right)\right] = 0$$

and multiplied by J

$$J \cdot \left(\frac{\partial U}{\partial \tau}\right) + \left(\frac{\partial y}{\partial \eta}\right) \cdot \left(\frac{\partial E}{\partial \xi}\right) - \left(\frac{\partial x}{\partial \eta}\right) \cdot \left(\frac{\partial F}{\partial \xi}\right) - \left(\frac{\partial y}{\partial \xi}\right) \cdot \left(\frac{\partial E}{\partial \eta}\right) + \left(\frac{\partial x}{\partial \xi}\right)$$

$$\cdot \left(\frac{\partial F}{\partial \eta}\right) = 0. \tag{4.14}$$

To obtain the conservative form, all terms must be brought under the derivative. The differential rules for this are

$$\frac{\partial}{\partial \tau}(J \cdot U) = J \cdot \left(\frac{\partial U}{\partial \tau}\right) + U \cdot \left(\frac{\partial J}{\partial \tau}\right)$$

$$\Rightarrow J \cdot \left(\frac{\partial U}{\partial \tau}\right) = \frac{\partial}{\partial \tau}(J \cdot U) - U \cdot \left(\frac{\partial J}{\partial \tau}\right),$$

$$\frac{\partial}{\partial \xi}\left(\frac{\partial y}{\partial \eta} \cdot E\right) = \left(\frac{\partial y}{\partial \eta}\right) \cdot \left(\frac{\partial E}{\partial \xi}\right) + E \cdot \frac{\partial}{\partial \xi}\left(\frac{\partial y}{\partial \eta}\right)$$

$$\Rightarrow \left(\frac{\partial y}{\partial \eta}\right) \cdot \left(\frac{\partial E}{\partial \xi}\right) = \frac{\partial}{\partial \xi}\left(\frac{\partial y}{\partial \eta} \cdot E\right) - E \cdot \frac{\partial}{\partial \xi}\left(\frac{\partial y}{\partial \eta}\right),$$

$$\frac{\partial}{\partial \xi}\left(\frac{\partial x}{\partial \eta} \cdot F\right) = \left(\frac{\partial x}{\partial \eta}\right) \cdot \left(\frac{\partial F}{\partial \xi}\right) + F \cdot \frac{\partial}{\partial \xi}\left(\frac{\partial x}{\partial \eta}\right)$$

$$\Rightarrow \left(\frac{\partial x}{\partial \eta}\right) \cdot \left(\frac{\partial F}{\partial \xi}\right) = \frac{\partial}{\partial \xi}\left(\frac{\partial x}{\partial \eta} \cdot F\right) - F \cdot \frac{\partial}{\partial \xi}\left(\frac{\partial x}{\partial \eta}\right),$$

$$\frac{\partial}{\partial \eta}\left(\frac{\partial y}{\partial \xi} \cdot E\right) = \left(\frac{\partial y}{\partial \xi}\right) \cdot \left(\frac{\partial E}{\partial \eta}\right) + E \cdot \frac{\partial}{\partial \eta}\left(\frac{\partial y}{\partial \xi}\right)$$

$$\Rightarrow \left(\frac{\partial y}{\partial \xi}\right) \cdot \left(\frac{\partial E}{\partial \eta}\right) = \frac{\partial}{\partial \eta}\left(\frac{\partial y}{\partial \xi} \cdot E\right) - E \cdot \frac{\partial}{\partial \eta}\left(\frac{\partial y}{\partial \xi}\right),$$

$$\frac{\partial}{\partial \eta}\left(\frac{\partial x}{\partial \xi} \cdot F\right) = \left(\frac{\partial x}{\partial \xi}\right) \cdot \left(\frac{\partial F}{\partial \eta}\right) + F \cdot \frac{\partial}{\partial \eta}\left(\frac{\partial x}{\partial \xi}\right)$$

$$\Rightarrow \left(\frac{\partial x}{\partial \xi}\right) \cdot \left(\frac{\partial F}{\partial \eta}\right) = \frac{\partial}{\partial \eta}\left(\frac{\partial x}{\partial \xi} \cdot F\right) - F \cdot \frac{\partial}{\partial \eta}\left(\frac{\partial x}{\partial \xi}\right).$$

Substituting this into Eq. 4.14 yields

$$\left[\frac{\partial}{\partial \tau}(J \cdot U) - U \cdot \frac{\partial J}{\partial \tau}\right] + \left[\frac{\partial}{\partial \xi}\left(\frac{\partial y}{\partial \eta} \cdot E\right) - E \cdot \frac{\partial}{\partial \xi}\left(\frac{\partial y}{\partial \eta}\right)\right]$$

$$- \left[\frac{\partial}{\partial \xi}\left(\frac{\partial x}{\partial \eta} \cdot F\right) - F \cdot \frac{\partial}{\partial \xi}\left(\frac{\partial x}{\partial \eta}\right)\right] - \left[\frac{\partial}{\partial \eta}\left(\frac{\partial y}{\partial \xi} \cdot E\right) - E \cdot \frac{\partial}{\partial \eta}\left(\frac{\partial y}{\partial \xi}\right)\right]$$

$$+ \left[\frac{\partial}{\partial \eta}\left(\frac{\partial x}{\partial \xi} \cdot F\right) - F \cdot \frac{\partial}{\partial \eta}\left(\frac{\partial x}{\partial \xi}\right)\right] = 0.$$

If the conservative and non-conservative terms are combined, it can be seen that the last term in square brackets becomes zero, since the derivatives of the metric derivatives in the other spatial directions and the time derivative of the Jacoby determinant vanish

$$
\frac{\partial}{\partial \tau}(J \cdot U) + \frac{\partial}{\partial \xi}\left(\frac{\partial y}{\partial \eta} \cdot E\right) - \frac{\partial}{\partial \xi}\left(\frac{\partial x}{\partial \eta} \cdot F\right) - \frac{\partial}{\partial \eta}\left(\frac{\partial y}{\partial \xi} \cdot E\right) + \frac{\partial}{\partial \eta}\left(\frac{\partial x}{\partial \xi} \cdot F\right)
$$
$$
- \underbrace{\left[U \cdot \frac{\partial J}{\partial \tau} + E \cdot \frac{\partial}{\partial \xi}\left(\frac{\partial y}{\partial \eta}\right) - F \cdot \frac{\partial}{\partial \xi}\left(\frac{\partial x}{\partial \eta}\right) - E \cdot \frac{\partial}{\partial \eta}\left(\frac{\partial y}{\partial \xi}\right) + F \cdot \frac{\partial}{\partial \eta}\left(\frac{\partial x}{\partial \xi}\right)\right]}_{0} = 0.
$$

Thus the conservative form of the conservation equations in curvilinear coordinates is given by

$$
\frac{\partial}{\partial \tau}\underbrace{(J \cdot U)}_{\hat{U}} + \frac{\partial}{\partial \xi}\underbrace{\left(\frac{\partial y}{\partial \eta} \cdot E - \frac{\partial x}{\partial \eta} \cdot F\right)}_{\hat{E}} + \frac{\partial}{\partial \eta}\underbrace{\left(-\frac{\partial y}{\partial \xi} \cdot E + \frac{\partial x}{\partial \xi} \cdot F\right)}_{\hat{F}} = 0
$$

respectively

$$
\frac{\partial}{\partial \tau}\hat{U} + \frac{\partial}{\partial \xi}\hat{E} + \frac{\partial}{\partial \eta}\hat{F} = 0 \tag{4.15}
$$

with the new vectors

$$
\hat{U} = J \cdot U, \tag{4.16}
$$

$$
\hat{E} = \frac{\partial y}{\partial \eta} \cdot E - \frac{\partial x}{\partial \eta} \cdot F, \tag{4.17}
$$

$$
\hat{F} = -\frac{\partial y}{\partial \xi} \cdot E + \frac{\partial x}{\partial \xi} \cdot F. \tag{4.18}
$$

Equation 4.15 is the form of the conservation equation to be discretized in curvilinear coordinates. The discretization of the partial derivatives and $\partial/\partial \xi$ is $\partial/\partial \eta$ done identically to and $\partial/\partial x$ $\partial/\partial y$ with central differences or the one-sided forward or backward differences.

4.3.4   Block Structured Meshe

Block-structured meshes are composed of several, usually structured mesh blocks. Figure 4.8 shows the schematic of a combined O-H mesh. An O mesh was placed around the geometry to optimally resolve the boundary layer. In the rest of the field an H-grid is used. This combination reduces the skewness of the mesh cells. The truncation errors become small and the accuracy of the numerical solution is improved.

It is important that the flow quantities at the block boundaries are passed correctly. If the net points of the two blocks are identical at the boundaries, the flow quantities can simply be passed from one boundary to the other. If the net points of the two blocks at the boundaries are not identical, the values of one block boundary must be interpolated to the net points of the other block boundary.

Figure 4.9 shows a block-structured O-H mesh around an airfoil generated with ICEM-CFD. The mesh consists of a total of nine blocks (eight H-mesh blocks 1–8 and one O-mesh block 9) around the airfoil. The points at the block boundaries between the mesh blocks coincide here, so that no interpolation is necessary.

Block-structured meshes are also frequently used for more complex geometries that are composed of several parts. A typical example is the flow calculation through one or more stages of a compressor or a turbine. Here, separate meshes are often generated for the impeller and the guide wheel, which are then assembled (Fig. 4.10).

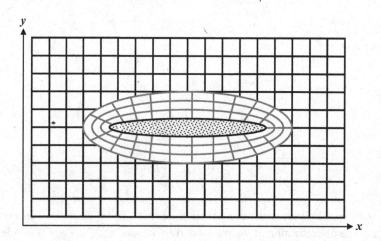

Fig. 4.8 Schematic of a block-structured O-H mesh (O mesh *red*, H mesh *black*)

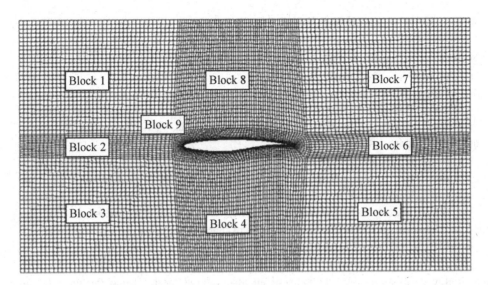

Fig. 4.9 Block-structured mesh around an aerofoil profile

Fig. 4.10 Block structured mesh around the blade profiles of a turbine stage. (Chima [9])

4.4   Unstructured Meshe

Section 4.3 dealt with structured meshs. Structured means that they have a regularity. The net lines of the same variables such as ξ = const. never overlap, neither do the lines and η = const. ζ = const. Each net point is assigned an increasing index i, j, k. The mesh cells are usually **hexahedra** (Fig. 4.11 left). The finite difference operators need structured meshes because, for example, they connect the mesh points $i - 1$, i and $i + 1$ in the case of the central spatial difference.

This structure is not necessary for the finite volume discretization. The calculation takes place directly in the physical space. A transformation into the computational space is not necessary. Therefore, unstructured meshes can also be used in finite volume methods. Unstructured meshes have a very high flexibility. They can be easily adapted to complex geometries consisting of several bodies and with sharp-edged corners. Here, the mesh cells are usually **tetrahedra** (Fig. 4.11 center) or **prisms** (Fig. 4.11 right), but they can also be other arbitrary solid elements.

Figure 4.12 shows the schematic of an unstructured mesh around the three bodies marked in red.

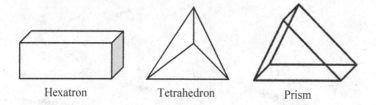

Hexatron   Tetrahedron   Prism

Fig. 4.11 Typical volume elements of a mesh

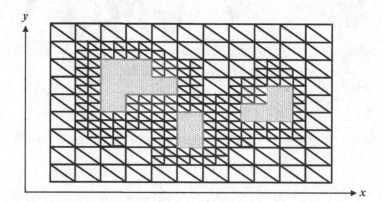

Fig. 4.12 Schematic of an unstructured mesh

Unstructured meshes can be easily adapted to areas with strong flow gradients such as boundary layers, wakes or compaction joints. Either existing cells are simply subdivided or the cells are moved into these areas.

A disadvantage of unstructured meshes is the more complex logistics of the mesh. For each volume element, the neighboring elements and points must be known to the program in order to be able to form the spatial differences.

4.5 Mesh Adaptation

4.5.1 Mesh Densification

At the boundary of the solid, the velocity is zero (adhesion condition) and a boundary layer profile for the velocity *u is established*, as shown in Fig. 4.13 on the left. This boundary layer profile must be resolved in order to be able to calculate the forces and moments and any detachments correctly. To do this, the mesh is compressed towards the solid boundary in the frictionally constrained calculation (Fig. 4.13 right). For good accuracy, the boundary layer should be resolved with at least ten mesh points normal to the surface, in this case in the *y-direction*. Its thickness can be estimated from theory before mesh compaction.

Modern mesh generation programs usually perform mesh compaction automatically at the solid boundary to resolve the boundary layer. Likewise, they automatically compact in areas where the geometry is highly curved, such as at kinks or edges like airfoil leading and trailing edges.

Figure 4.14 shows such a mesh condensed at the wall and at the leading and trailing edge for an airfoil. To resolve the wall boundary layer, a structured O mesh with rectangular faces is used here. This reduces the discretization errors and the accuracy of the

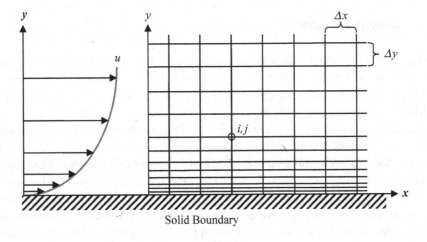

Fig. 4.13 Mesh compaction at the solid boundary

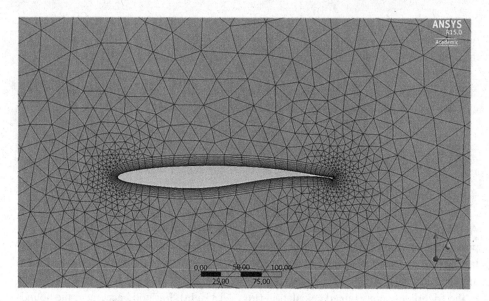

Fig. 4.14 Compacted unstructured mesh around an airfoil with structured O-mesh in the boundary layer

numerical solution is found to be better than using an unstructured mesh in the boundary layer. This is due to the fact that the skewness is smaller for rectangular surfaces than for triangular surfaces.

Also clearly visible are the mesh compressions in the leading and trailing edge areas. The stronger curvature of the geometry results in stronger flow gradients, which must be resolved in order to obtain good accuracy.

4.5.2 Adaptive Meshe

In order to be able to resolve flow gradients such as the flow boundary layer well, many mesh points must lie in the boundary layer. Since the approximate position of the boundary layer is already known before the calculation, the mesh can already be compressed at the solid boundary during the mesh generation. It becomes more difficult if strong flow gradients arise only during the calculation. For example, the position of compaction joints is not yet known at the beginning. Here it is advantageous if the flow calculation program recognizes the position of shocks and automatically compresses the mesh at these points.

Through these so-called adaptive meshes, the computation can be performed accurately and efficiently. In areas with low flow gradients, the mesh is thinned to save computation time and in areas with strong flow gradients, the mesh is compacted to achieve good accuracy. Mesh compaction is usually controlled by a flow gradient such as static pressure.

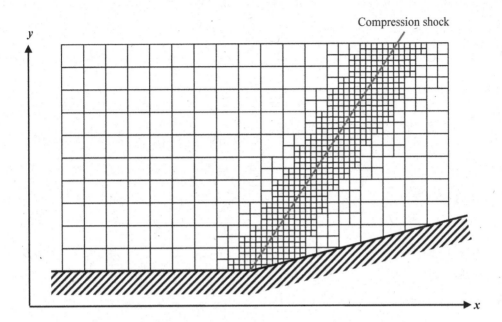

Fig. 4.15 Schematic of an adapted mesh around an inclined compaction joint

Figure 4.15 shows a Cartesian mesh with a mesh compression at the compression joint (red line). In the remaining flow field, where the flow gradients are small, the mesh is coarser. Thus, a good accuracy results at the joint and a shorter computation time than with constant high resolution in the entire domain (at the solid body surface for the resolution of the boundary layer was omitted here for reasons of clarity).

Solution Methods

5

5.1   Aim of This Chapter

How can the differential equations, i.e. the discretized differential equations, now be solved on the computational mesh? There are numerous solutifon algorithms for this, which became more and more exact and faster in the course of time and with increasing computer performance. The development went here from the central schemes over the Upwind schemes to the High-Resolution schemes.

While the classical central methods provide good results for subsonic flows, they often have problems for supersonic and hypersonic flows with strong shocks, such as those occurring during the re-entry of a spacecraft into the atmosphere. They converge poorly and compute the compression shocks too inaccurately. In contrast, the so-called upwind methods are very stable and accurate in the area of the compression shock, but too inaccurate in the rest of the flow field. Only with the development of the modern, so-called high-resolution methods could both advantages be combined. They achieve a good stability and a good accuracy in the entire flow area for both shock-free and shocked flows.

These solution methods together with their advantages and disadvantages are the content of this chapter. After reading, you should be able to answer the following questions:

1. What are the three classes of solution schemes?
2. What is the difference between these three classes?
3. Name three groups of solution methods that use central spatial discretization.
4. Name one advantage and disadvantage of each of the upwind methods.
5. Which property is satisfied by a monotone solution method?

© The Author(s), under exclusive license to Springer Fachmedien Wiesbaden GmbH, part of Springer Nature 2022
S. Lecheler, *Computational Fluid Dynamics*,
https://doi.org/10.1007/978-3-658-38453-1_5

6. Can a method with second order spatial accuracy be monotonic?
7. Why was the TVD condition introduced?
8. What additional condition must a TVD process meet in order to obtain a physical solution?
9. What are limiter functions good for?
10. What are the names of the methods that provide accurate solutions for both continuous and discontinuous flow variables?

5.2   Overview

Three classes of solution methods are distinguished (Fig. 5.1):

- The **central methods** with central spatial discretization with a spatial accuracy of second order.
- The **upwind methods of** one-sided spatial discretization and first order spatial accuracy.
- The **high-resolution methods** with one-sided or central spatial discretization and a spatial accuracy of second order. They are also called TVD

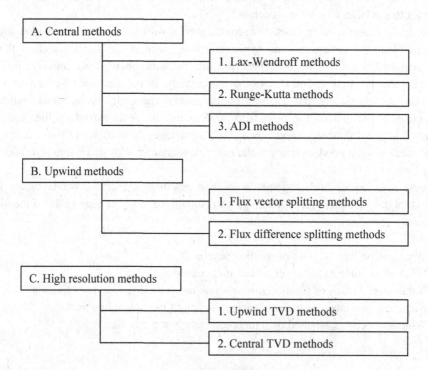

Fig. 5.1 Overview of the three classes of solution methods

What these terms mean is explained in the next chapters for the respective schemes.

5.3   Central Methods

5.3.1   Overview

Central methods are characterized by a central discretization of the spatial derivatives. This leads to a good spatial accuracy of second order (cf. Sect. 3.3). These methods often have convergence problems in the case of strong compression shocks. Figure 5.2 shows an overview of the most important central methods and their namesakes.

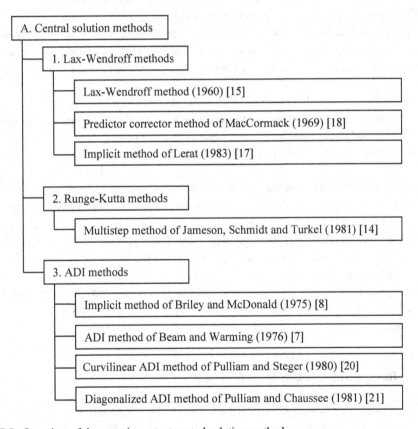

Fig. 5.2 Overview of the most important central solution methods

5.3.2 Lax-Wendroff Method

Between 1957 and 1964, Peter Lax and Burton Wendroff developed a solution method for the unsteady Euler and Navier-Stokes equations, which became a milestone for modern computational fluid dynamics [15]. It was the first centralized method for solving the Euler equations. In the meantime, there are numerous variants, which, however, have the following properties in common:

- The spatial discretization is done centrally.
- Lax-Wendroff methods are of second order accuracy, both in time and in space. They can therefore be used for both time-accurate and time-asymptotic solutions.
- The solution is combined for the temporal and spatial derivatives. However, this has the disadvantage that the stationary solution depends on the selected time step.
- The temporal discretization is usually explicit, but there are also implicit variants.
- Lax-Wendroff methods are stable even without an additive numerical viscosity due to their combined second order temporal and spatial discretization. However, oscillations arise in the solution at discontinuity points, e.g., compression shocks.

In the classical Lax-Wendroff method, the second derivatives of the Taylor series expansion are also discretized. This means that the algebra is relatively complex and therefore mistakes can easily be made during programming. Because of this, numerous variants have been developed over the years. The best known are:

- The Predictor-Corrector method by Robert W. MacCormack [18]. It is much simpler and easier to program. In the predictor step, first order forward differences are used, which are actually unstable for supersonic regions. In the subsequent corrector step, first order backward differences are used, which are actually unstable for subsonic flows. Nevertheless, the combination is stable and even of second order accuracy, since the truncation errors of both steps cancel.
- The implicit Lax-Wendroff method of Alain Lerat [17] is also stable for larger time steps or CFL numbers and resolves compaction shocks very well even without additive numerical viscosity.

5.3.3 Runge-Kutta Multi-Step Method

In contrast to the Lax-Wendroff methods, the Runge-Kutta multistep methods perform the time solution independently of the spatial discretization. This has the advantage that the stationary solution is independent of the time step. This method was introduced by Antony Jameson, W. Schmidt and E. Turkel in 1981 for finite volume-method which is very efficient and accurate [14]. The main properties are:

- Central spatial differences with second order accuracy.
- The explicit temporal integration is usually carried out via four intermediate steps. The temporal accuracy can be chosen from first to fourth order, depending on the chosen pre-factors of the intermediate steps.

In the Runge-Kutta multistep method, numerous measures can be used to accelerate convergence when only time-asymptotic solutions are sought. Thus, stationary solutions can be obtained much faster than with purely explicit methods. The most common measures are:

- With local time step control, a separate, local time step is used for each network cell, which fulfils the local CFL condition of the network cell. Large time steps can therefore be used for large network cells. Although each cell then has its own time level, this is not important for the steady-state solution.
- In enthalpy damping, the energy equation is additionally damped by adding an enthalpy term, which disappears again in the convergent solution.
- Residual smoothing allows for larger CFL numbers than purely explicit methods and thus moves towards implicit temporal discretization. In residual smoothing, the high-frequency disturbances in the residual are smoothed.
- The multigrid technique uses solutions on nets of different fineness. The solution is transferred from a finer mesh to a coarser mesh. This accelerates convergence to the steady-state solution because larger time steps can be used on coarser meshes. The coarse solution is then interpolated back onto the fine computational mesh. This eliminates the high-frequency errors more quickly.

5.3.4 ADI Method

In the ADI methods, as in the Runge-Kutta methods, the temporal solution is performed independently of the spatial discretization, which means that the stationary solution is independent of the time step. The ADI (Alternating Direction Implicit) methods were developed in 1975 by W. R. Briley and Henry McDonald [8] and in 1976 by Richard M. Beam and Robert F. Warming [7] and have the following general properties:

- Due to the implicit temporal discretization, much larger time steps or CFL numbers can be used. However, due to the implicit formulation, a system of equations must be solved for all mesh points. This increases the computational effort per iteration, but reduces the necessary number of iterations until convergence considerably.
- In order to solve the system of equations efficiently, each spatial direction is solved separately on the implicit side. This results in three one-dimensional systems of equations that have to be solved by means of matrix inversion algorithms. Due to the central spatial discretization, the one-dimensional systems of equations are only

occupied on the three central diagonals and are referred to as a tridiagonal matrix (see Eq. 3.30). Efficient inversion algorithms exist for them.

- For increased stability and better shock resolution, a numerical viscosity is also added on the explicit side. It usually contains second order terms for shock resolution and fourth order terms for damping in continuous regions. By means of a pressure gradient switch, the second order additive viscosity is activated at the shock. In addition, a second added on the implicit side to improve stability and allow larger time steps.

There are also other variants of ADI methods:

- Thomas H. Pulliam and Joseph L. Steger transformed the ADI method into oblique coordinates and developed powerful programs for solving the thin-layer and Reynolds-averaged Navier-Stokes equations [20].
- In the diagonal variant of Thomas H. Pulliam and Denny S. Chaussee [21], the block-tridiagonal matrices (5 × 5 in 3D, 4 × 4 in 2D, 3 × 3 in 1D) are transformed on the implicit side so that only one scalar size remains. This considerably reduces the computational effort for the matrix inversion, but for stability reasons the CFL numbers must also be reduced again.
- The author himself developed a variant of the ADI-method with implicit characteristic boundary conditions during his PhD [16]. Due to the characteristic boundary conditions, a good accuracy at the edges is achieved, while the implicit formulation of the boundary conditions allows large CFL numbers or time steps and thus enables short computation times.

5.4   Upwind Methods

5.4.1   Overview

The solution methods in the previous chapter are based on the central spatial discretization. It links mesh points from all spatial directions without considering the physical propagation direction of disturbances. They are based on the assumption that the flow variables are continuous and can be approximated by a Taylor series expansion.

Since this assumption no longer applies at discontinuity points such as compression joints, oscillations occur there or the process can become completely unstable. Therefore, a so-called numerical or artificial viscosity must be added. Only then are central processes for non-linear processes such as compression shocks stable and oscillations in the shock environment are significantly reduced.

How much numerical damping must be added, however, depends on the flow problem and is specified by the user via an input value. The numerical damping should be large enough to ensure good convergence, but small enough not to falsify the solution too much.

Therefore, this input value should be adapted to the respective flow problem, e.g. it should be higher for the occurrence of strong discontinuities than for pure subsonic flows.

In order to reduce this uncertainty about the size of the input value, Richard Courant, E. Isaacson and M. Reeves [10] came up with the idea of developing the so-called upwind methods as early as 1952. These already take into account the physical processes in the flow, namely the propagation direction of disturbances along the characteristics, during the discretization. Upwind methods therefore have an inherent physical damping which automatically adapts to the flow gradients. For strong shocks it becomes larger than for pure subsonic flow. The addition of a numerical damping as in central methods is not necessary and the arbitrariness in the size of the input value is eliminated.

Upwind methods are therefore very stable solution methods. Compression shocks are resolved without oscillations. However, the pure upwind methods only use first order forward and backward differences, so that they only have a spatial accuracy of first order. Usually, however, second order accuracy is required for technical applications. This can be remedied by second order upwind methods, also known as high resolution or TVD methods. These are discussed in Sect. 5.5.

Figure 5.3 shows an overview of the most important upwind processes.

Two general characteristics of upwind methods should be noted by way of introduction:

- Upwind methods are **monotonic methods**. This means that they do not allow or generate new extrema and unphysical discontinuities. This property prevents oscillations and unphysical solutions. Unfortunately, monotone methods are accurate at most of first order.
- Upwind first order methods correspond to central methods with a second order additive viscosity. One can mathematically transform them into each other. However, this also means that even methods with central spatial discretization and a second order numerical viscosity are only first order accurate (like the Upwind methods). Only when a fourth order numerical viscosity is added do central methods become second order accurate. However, they then again have the problems with oscillations at the joint.

There are numerous variants of first order upwind methods, but they can all be classified as the following two methods:

- The first order **flux vector splitting methods.** In them, the spatial flux terms are split according to the sign of the eigenvalues.
- The first order **flux difference splitting method** with the Riemann solver. They are also known as the Godunov method after Sergei K. Godunov, who proposed this method in 1959. They are more accurate but also more complex than the flux vector splitting methods, since they solve the so-called shock wave or Riemann problem at each cell boundary.

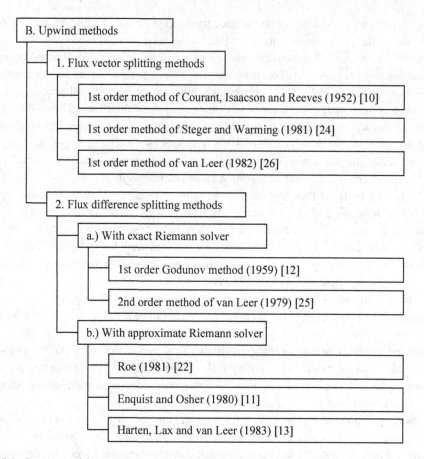

Fig. 5.3 Overview of the most important upwind solution methods

5.4.2 Flux Vector Splitting Method

In flux vector splitting methods, the flux terms are split according to the sign of the eigenvalues. The one-dimensional conservation equation

$$\frac{\partial U}{\partial t} + \frac{\partial E}{\partial x} = 0 \tag{5.1}$$

then reads, for example

$$\frac{\partial U}{\partial t} + \frac{\partial E^+}{\partial x} + \frac{\partial E^-}{\partial x} = 0. \tag{5.2}$$

- $\frac{\partial E^+}{\partial x}$ contains only the terms with positive eigenvalues, that is, the downstream characteristics or information $(u, u + a)$, where is u the flow velocity and a the sound velocity. These are discretized with a backward difference.
- $\frac{\partial E^-}{\partial x}$ contains only the terms with negative eigenvalues, i.e. the upstream characteristics or information $(u - a)$. These are discretized with a forward difference.

Splitting according to the signs of the eigenvalues has the disadvantage that the exact formulation is no longer conservative. This would mean that compaction shocks would no longer be automatically detected and correctly calculated. Therefore, numerous measures have been taken to both preserve conservatism and to account for the direction of propagation of the information (characteristics) as accurately as possible. The best known schemes are

- the flux vector splitting method of Joseph L. Steger and Robert F. Warming [24].
- the flux vector splitting method by Bram van Leer [26].

In Sect. 5.6 the solutions of all methods mentioned in this chapter are compared.

5.4.3 Flux Difference Splitting Method

The class of so-called flux-difference splitting methods is based on a method of Sergei K. Godunov, which states that the solution is constant in the cell, but jumps at the cell edge according to the exact Euler solution [12]. Therefore, the flux-difference splitting methods are also called Godunov-methods. At the cell edges, the so-called shock wave – or Riemann – problem has to be solved with the expansion fan, the contact area and the shock wave. Figures 5.4, 5.5 and 5.6 show the three steps for the calculation of the u_i^{n+1} next time step for a so-called cell center method, in which the flow variables are stored in the cell center.

In the first step, the flow variables u_{i-1}^n, u_i^n, u_{i+1}^n etc. are constant in all cells. At the cell edges $i - \frac{1}{2}$, $i + \frac{1}{2}$ etc. jumps or discontinuities occur (Fig. 5.4).

In the second step, the so-called Riemann problem is solved at all cell edges $i - \frac{1}{2}$, etc.. $i + \frac{1}{2}$ It describes the flow at discontinuity points by means of an expansion fan, a contact area and a compression impact. This can be done exactly (exact Riemann solver) or approximated (approximated Riemann solver) to save computation time. This results in new values for the left and right cell edge (Fig. 5.5).

In the third step, the states from the left and right cell edges are averaged to again obtain constant flow variables at the new time $n + 1$ in each cell (Fig. 5.6).

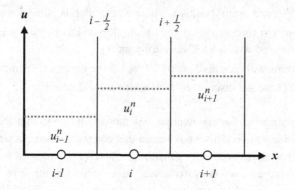

Step 1:
Constant
distribution
within each
cell at time
step n

Fig. 5.4 First step of the flux difference splitting method

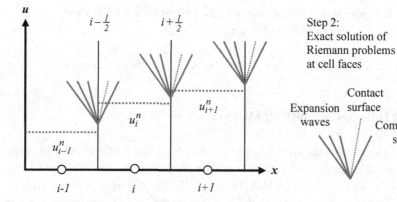

Step 2:
Exact solution of
Riemann problems
at cell faces

Contact
Expansion surface
waves
Compression
shock

Fig. 5.5 Second step of the flux difference splitting method

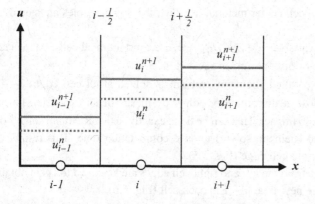

Step 3:
Averaging of the
values from left and
right cell faces at
time step $n+1$

Fig. 5.6 Third step of the flux difference splitting method

By solving the Riemann problem, flux-difference splitting methods involve a lot of physics:

- they are conservative, i.e. Shocks are detected automatically.
- they are very stable and monotonic, i.e. they avoid new extremes and oscillations
- they have a very good impact resolution.
- unfortunately, they too are only first order accurate.

There are also several variants of flux-difference splitting methods:

- the Godunov method with an exact Riemann solver [12].
- the Enquist-Osher method with an approximated Riemann solver [11].
- the Roe method also with an approximated Riemann-solver [22].

Calculation results and comparisons with the other solution methods are shown in Sect. 5.6.

5.4.4 Summary

All first order upwind methods can be formulated in a uniform discrete notation:

$$U_i^{n+1} = U_i^n - \frac{\Delta t}{\Delta x} \cdot \left(E_{i+\frac{1}{2}}^* - E_{i-\frac{1}{2}}^* \right) \tag{5.3}$$

with the flux terms E at the cell walls and $i + \frac{1}{2} i - \frac{1}{2}$ as a function of the maintenance vector U at the cell centers $i - 1$, and $ii + 1$

$$E_{i+\frac{1}{2}}^* = f(U_i, U_{i+1}) \quad \text{and} \quad E_{i-\frac{1}{2}}^* = f(U_{i-1}, U_i). \tag{5.4}$$

The flow terms are for the different methods:

- flux vector splitting scheme

$$E_{i+\frac{1}{2}}^* = E^-(U_{i+1}) + E^+(U_i), \tag{5.5}$$

- Godunov scheme

$$E_{i+\frac{1}{2}}^* = E\left[U_{i+\frac{1}{2}}^{(R)}(0, \ U_i, \ U_{i+1}) \right], \tag{5.6}$$

• Osher scheme

$$E^*_{i+\frac{1}{2}} = \frac{1}{2} \cdot (E_i + E_{i+1}) - \frac{1}{2} \cdot \sum_j \int_{\Gamma(j)} |\lambda_{(j)}| \cdot r^{(j)} \cdot dw, \tag{5.7}$$

• Roe scheme

$$E^*_{i+\frac{1}{2}} = \frac{1}{2} \cdot (E_i + E_{i+1}) - \frac{1}{2} \cdot \sum_j |\bar{\lambda}_{(j)}| \partial w_j \cdot \bar{r}^{(j)}, \tag{5.8}$$

with λ as eigenvalue, r as eigenvector and w as characteristic variable of the wave propagation. The details and formulas can be found in further literature such as [3], as they are beyond the scope of this book.

5.5   High-Resolution Methods

5.5.1   Overview

Central methods are second order accurate, but lead to oscillations in the solution at discontinuities, which lead to instability without numerical damping. Upwind methods, on the other hand, are very stable but unfortunately only accurate to first order. High-resolution or TVD methods, which are accurate to the second order and prevent the occurrence of oscillations at discontinuities, provide a remedy here. Figure 5.7 gives an overview of the most important methods of this class.

5.5.2   Monotonicity, TVD and Entropy Condition

The upwind methods presented in Sect. 5.4 have the advantage that they achieve a good shock resolution without oscillations. However, their accuracy is only of first order. To achieve second order accuracy, a linear distribution can be used instead of the constant distribution within the cell (see Figs. 5.4 and 5.8).

However, it turns out that the solutions of the second order upwind methods formed in this way also exhibit oscillations at discontinuity points like the second order central methods. The monotonicity condition could provide a remedy here, since it does not allow oscillating solutions. Unfortunately, only first order methods satisfy this monotonicity condition. For second order methods, however, it can be replaced by two similar conditions:

Fig. 5.7 Overview of the most important high-resolution solving methods

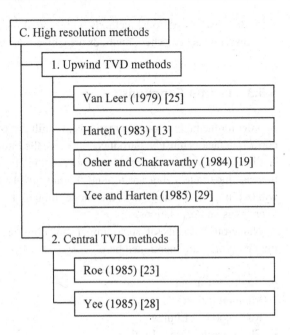

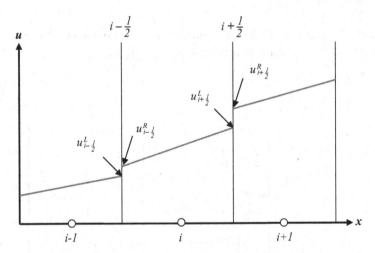

Fig. 5.8 Linear distribution in the cells as the basis of the high-resolution methods

- The **TVD condition** (Total Variation Diminishing) states that the total variation of a solution must decrease during the time iteration. It is a weakened monotonicity condition for methods of second order accuracy and also leads to oscillation-free solutions. However, the TVD condition allows unphysical solutions.
- To prevent this, the **entropy condition** is also introduced. It ensures that the solution is physical, i.e. the entropy must not decrease.

With these two conditions, methods with second order accuracy and without oscillations can now be achieved. These methods are called high-resolution methods or TVD methods.

5.5.3 Limiter Functions

In order for the high-resolution methods to fulfill the TVD and entropy condition, nonlinear correction factors must be introduced, the so-called limiters. They ensure that, for example, at discontinuity points such as compression joints, the slopes of the cell sizes do not increase indefinitely, but are limited. Figure 5.9 shows i a linear distribution in the cell without limiter (dashed) and the slope limited by the limiter (solid). This prevents overshoots and oscillations.

Numerous variants of such limiter functions are given in the literature. The most important ones are mentioned here:

- Van Leer Limiter
- Minmod limiter
- Roe Superbee Limiter
- Chakravarthy-Osher Limiter.

In the examples in Sect. 5.6, for example, solutions with the Minmod and Superbee limiters are shown. Details can be found in the literature, e.g. [3].

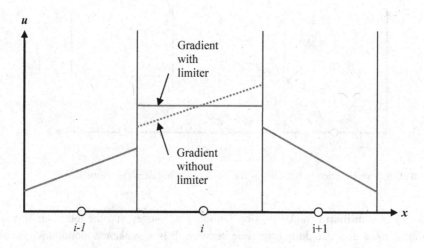

Fig. 5.9 Limiting the gradients at turning points by limiters

5.5.4 Summary

Finally, the previous procedure for the generation of TVD methods with the upwind discretization is summarized again:

- Basis is a monotone upwind method of first order accuracy.
- Extension of the method to second order accuracy, e.g. by a linear distribution within the grid cells instead of the constant distribution.
- Preventing the occurrence of oscillations by fulfilling the TVD condition, i.e. the amplitude of gradients is limited by limiters.
- Avoidance of unphysical solutions by checking the entropy condition for the second order TVD method with the limiters. This means that solutions where the entropy decreases are excluded.

Figure 5.10 shows the most important properties of these three classes of solution methods. They will be explained in more detail in the next chapters.

An interesting result in the development of TVD methods for upwind discretization is that the TVD principle can also be applied to methods with central spatial discretization [28]. This results in a nonlinear numerical viscosity without empirical input values, which also leads to solutions of second order accuracy without oscillations. An advantage of the central TVD methods is that existing solution methods with central spatial differences such as the Lax-Wendroff or Runge-Kutta methods can be used. Only the additive numerical viscosity has to be formulated in such a way that it satisfies the TVD and the entropy condition.

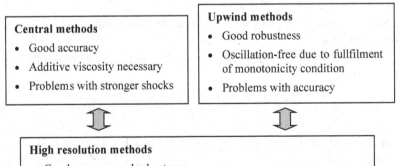

Fig. 5.10 Overview of the most important properties of the three classes of solution methods

5.6   Comparison of the Methods

In the following, results of the main solution methods are presented and compared using two examples taken from [3]:

- The steady flow through a divergent nozzle. This is about the resolution of the compression shock.
- The unsteady flow in a shock wave tube after approx. 6 ms. This more complex flow contains three discontinuities, a compression shock, an expansion fan and a contact discontinuity.

5.6.1   Stationary Flow Through a Divergent Nozzle

This example is characterized by the following dimensionless properties:

- Divergent nozzle with cross-sectional area ratio $A(x) = 1, 398 + 0, 347 \cdot \tanh [0, 8 \cdot (x - 4)]$ for $0 \leq x \leq 10$
- Narrowest cross-sectional area $A^* = 0, 8$
- Supersonic inflow
- Stationary transonic flow with a compression shock at $x = 4$.

There is a theoretical solution for this flow, which is shown as a red line in Figs. 5.11–5.13. Three curves are shown along the dimensionless nozzle contour from $x = 0$ to $x = 10$ with the discontinuity (compression shock) at $x = 4$:

- The Mach number as a typical flow quantity, which is discontinuous at the shock and changes from supersonic to subsonic.
- The entropy related to the gas constant R as a typical measure of losses that increase abruptly at the shock.
- The mass flow error as a quantity for the satisfaction of the equation of conservation of mass.

The numerical solutions the following methods are compared with this theoretical solution:

1. Second order central method according to MacCormack without numerical damping (Fig. 5.11a)
2. Second order central method according to MacCormack with numerical damping according to MacCormack and Baldwin (Fig. 5.11b)
3. First order upwind method with flux vector splitting according to Steger and Warming (Fig. 5.12a)
4. First order upwind method with flux vector splitting according to Van Leer (Fig. 5.12b)

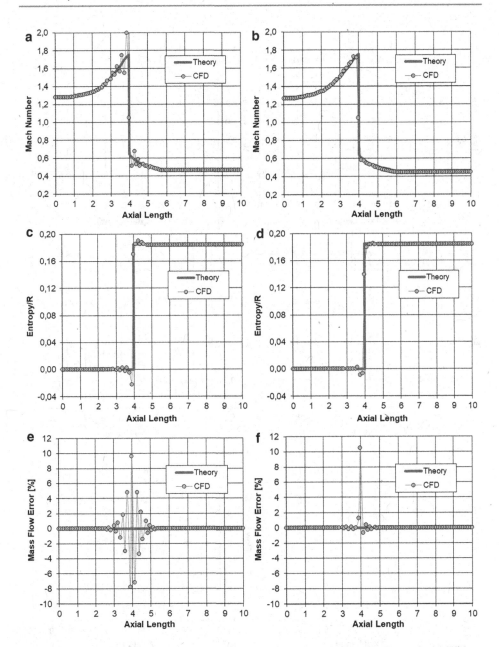

Fig. 5.11 Supersonic nozzle: second order central method according to MacCormack. (**a**) Without numerical damping, (**b**) with numerical damping

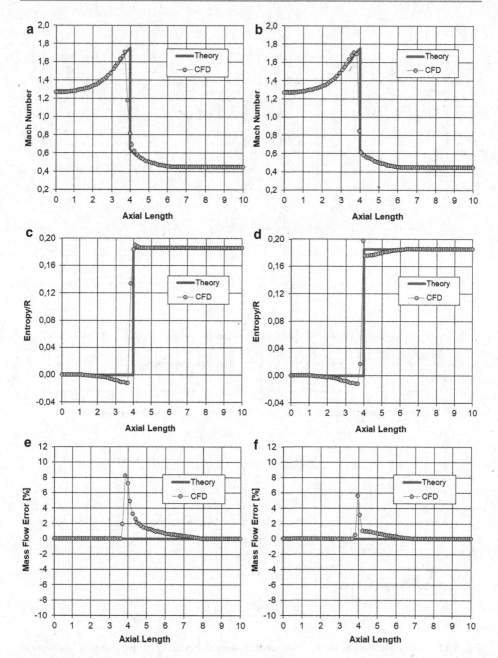

Fig. 5.12 Supersonic nozzle: first order upwind method with flux vector splitting according to Steger and Warming (**a**) and according to Van Leer (**b**)

5. First order upwind method with flux difference splitting according to Roe (Fig. 5.13a)
6. High-resolution second order method with minmod limiter according to Roe (Fig. 5.13b)

The numerical solutions are shown as a black line with circles. The first two solutions in Fig. 5.11a, b were obtained using MacCormack's central method. It is second order accurate, but is prone to unstable behavior and oscillations. These can be greatly reduced by suitable numerical damping, resulting in good numerical solutions:

- Figure 5.11a shows the numerical solution without numerical damping. The undershoots and overshoots before and after the impact are clearly visible. The discontinuous change of the Mach number at the impact is only inadequately captured. Significant deviations and oscillations are also visible in the entropy at the shock. The mass flow error fluctuates very strongly in the shock range.
- The addition of a suitable numerical damping improves the results significantly (Fig. 5.11b). For the resolution of the shock only three mesh points are needed and the peak at the beginning of the shock is well captured. The entropy still shows small undershoots before the shock. For the mass flow error, the oscillations at the shock become smaller. However, the error of 10% at the start of the surge remains.

The three solutions in Figs. 5.12a, b and in Fig. 5.13a were obtained using upwind methods. These are only accurate to first order, but their monotonicity property prevents oscillations as they occur:

- In the flux vector splitting method according to Steger and Warming, the oscillations at the joint have disappeared (Fig. 5.12a). However, the solution is inaccurate in the shock environment. Both the Mach number and entropy deviate significantly from the exact solution. The compactification shock is smeared and resolved with at least five points. This shows that this method dampens too much.
- The flux vector splitting method according to Van Leer (Fig. 5.12b) is somewhat better. Above all, the resolution of the compaction impact is clearly better with three points. The Mach number plot agrees well with the exact solution. Only the entropy deviates at the joint. This upwind method is less dissipative (damping) than the flux vector splitting method of Steger and Warming.
- A further improvement is obtained with the flux-difference splitting method according to Roe (Fig. 5.13a). The impact is very well resolved. Only the entropy deviates from the exact solution at the joint.

As expected, the most accurate solution can be obtained with the high-resolution method according to Roe (Fig. 5.13b). It is second order accurate, but without oscillations, since it uses a min-mod limiter and thus satisfies the TVD condition:

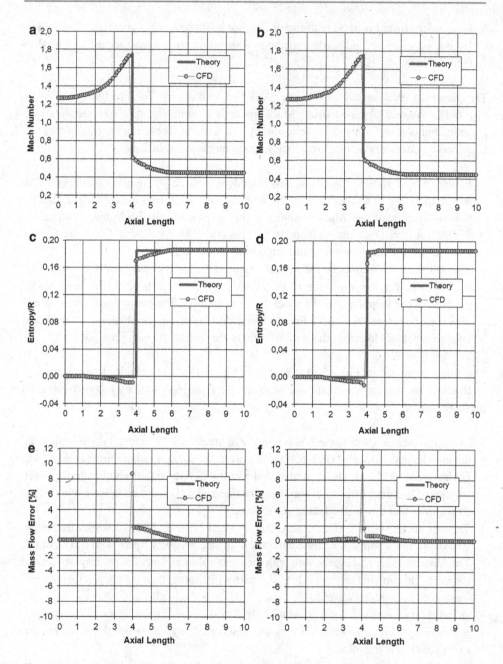

Fig. 5.13 Supersonic nozzle: first order upwind method with flux difference splitting according to Roe (**a**) and second order high-resolution method with minmod limiter according to Roe (**b**)

- The Mach number curve agrees very closely with the theory. The impact is recorded with three grid points. However, the entropy still shows deviations before and after the impact and the mass flow error is small except for the overshoot at the impact.

5.6.2 Unsteady Flow in a Shock Wave Tube

A shock wave tube is a tube filled with gas, which is divided by a membrane at the point x_0. At time $t = 0$ there are two states (Fig. 5.14):

- The left state L for $x < x_0$ with high pressure $p_L = 10^5$ and high density $\rho_L = 1$.
- The right state R for $x > x_0$ with a pressure $p_R = 10^4$, which is lower by a factor of 10, and the density $\rho_R = 0.125$, which is eight times lower.

The velocities at time $t = 0$ on both sides are $u_L = u_R = 0$. All quantities are dimensionless.

At time $t > 0$ (Fig. 5.14) the membrane bursts and the left high pressure region spreads into the right low pressure region. For $t > 0$, a shock wave and a contact area arise on the right side. At them the flow variables jump discontinuously. At the same time, an expansion fan spreads out on the left side at which the flow variables change continuously.

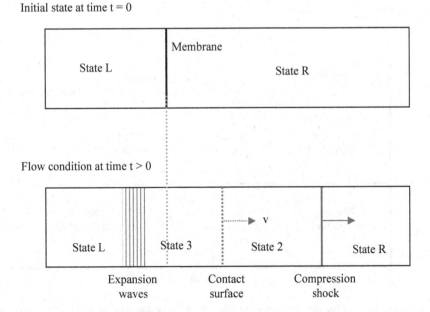

Fig. 5.14 Schematic representation of the shock wave problem

The propagating unsteady flow in this shock wave tube can be calculated exactly under the assumption that the viscous effects are small and that the tube is infinitely long so that no reflections occur at the two tube ends. The shock and the contact area propagate with constant velocities into regions with constant states.

Figure 5.15 shows the propagating characteristics, the two discontinuities impact and contact area as well as the expansion fan in the *x-t-diagram*. A distinction can be made between the following areas:

- Rightmost state R contains the undisturbed gas with the low pressure p_R.
- It is separated by a compression shock from state 2, which contains the already disturbed gas at low pressure.
- The contact surface separates state 2 from state 3 with high pressure, which has already been disturbed by the expansion fan.
- State 5 prevails in the expansion fan, in which the flow variables vary continuously.
- On the far left there is still the undisturbed high pressure area with condition L.

This is a challenging test case for the numerical methods, since they have to resolve both the shock wave and contact area with their discontinuous changes and the expansion fan with its continuous change of the flow variables.

For the following solution methods, the numerical solutions are compared with the theoretical solution:

1. Second order central method according to MacCormack without numerical damping (Fig. 5.16a)

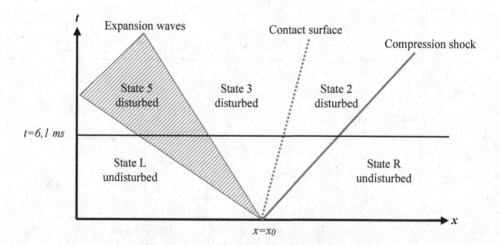

Fig. 5.15 *x-t-diagram* for the shock wave problem with the propagation of the compression shock, the contact area and the expansion fan

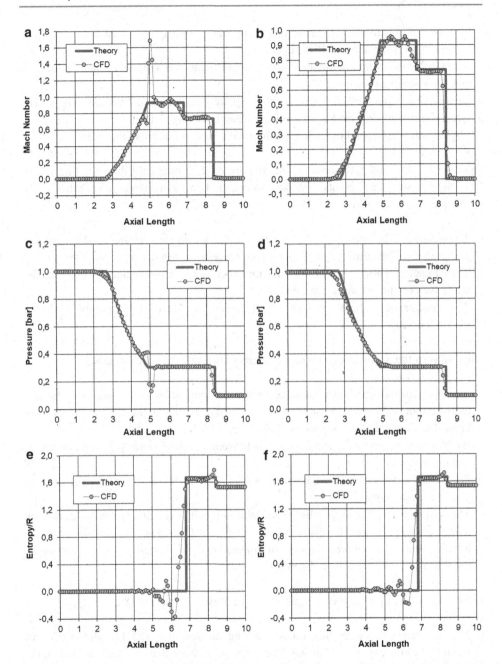

Fig. 5.16 Shock wave problem: Second order central method according to MacCormack. (**a**) Without numerical damping, (**b**) with numerical damping

2. Second order central method according to MacCormack with numerical damping according to Neumann and Richtmyer (Fig. 5.16b)
3. Central Second order ADI method according to Beam and Warming with explicit and implicit numerical damping (Fig. 5.17a)
4. First order upwind method with flux vector splitting according to Van Leer (Fig. 5.17b)
5. Second order upwind method according to Van Leer (Fig. 5.18a)
6. High-resolution Second order Roe method with Superbee limiter (Fig. 5.18b)

Figures 5.16, 5.17 and 5.18 show as a red line the exact theoretical solution at time $t = 6$, 1ms. Shown are Mach number, static pressure and the entropy/gas constant. Clearly visible are:

- The expansion fan between $2, 7 < x < 4, 9$ with a steady progression of Mach number and pressure. The entropy remains constant, since no losses occur.
- The contact area at $x \approx 6, 8$. Here, Mach number and entropy change discontinuously, while the pressure remains constant.
- The compression shock at $x \approx 8, 4$. Here, all three flow variables change discontinuously.

The results of the 3 s order central methods are shown in Figs. 5.16a, b and 5.17a:

- The MacCormack method without numerical damping shows very large undershoots and overshoots at the expansion fan and a smeared resolution of the discontinuities at the contact surface and the compression joint (Fig. 5.16a).
- By adding a numerical damping according to Neumann-Richtmyer [27], the overshoots and undershoots at the expansion fan have disappeared (Fig. 5.16b), but they remain at the contact surface and at the compression joint. The shock resolution is rather smeared with six mesh points.
- The ADI method according to Beam and Warming with explicit and implicit numerical damping shows a too dissipative solution (Fig. 5.17a). The discontinuities at the interface and at the compression joint are smeared over many mesh points. Nevertheless, an overshoot is still visible at the expansion fan.

The results of two upwind methods are shown in Figs. 5.17b and 5.18a:

- The first order Van Leer flux vector splitting method in Fig. 5.17b shows a better resolution of the expansion fan and the compression shock than the ADI method on the left side of Fig. 5.17. However, the contact area is still rather smeared.

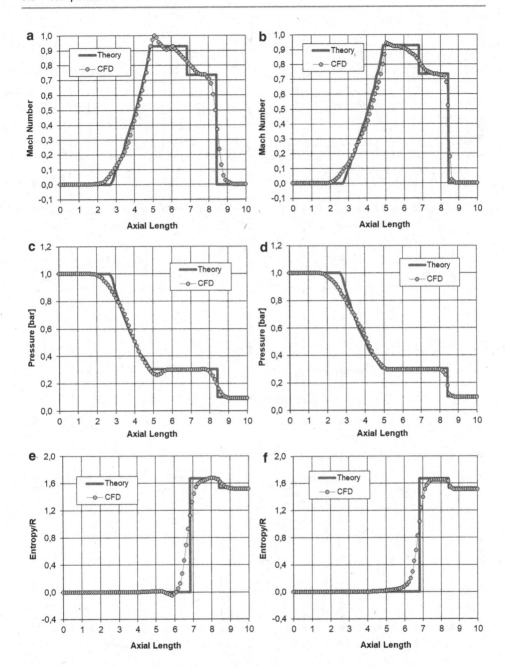

Fig. 5.17 Shock wave problem: Second order central ADI method according to Beam and Warming with explicit and implicit numerical damping (**a**) and first order upwind method with flux vector splitting according to Van Leer (**b**)

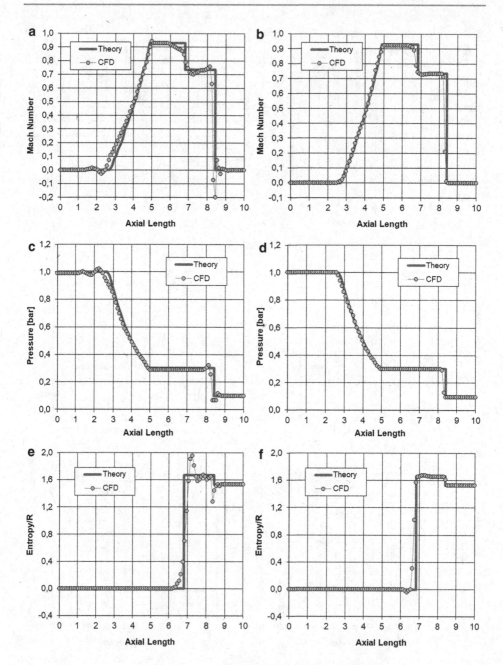

Fig. 5.18 Shock wave problem: Second order upwind method with flux splitting according to Van Leer (**a**) and second order high-resolution method according to Roe (**b**)

- A better resolution of the expansion fan is obtained with the second order flux splitting method according to Van Leer (Fig. 5.18a). However, due to the higher order of accuracy, overshoots and undershoots arise again at all three flow phenomena.

As expected, the best solution is again shown by the high-resolution method in Fig. 5.18b:

- It shows good resolution of the flow gradients since it is second order accurate. In addition, overshoots or undershoots are prevented at the onset by the TVD condition. and the Superbee limiter.

Typical Workflow of a Numerical Flow Calculation

6

6.1 Aim of This Chapter

This chapter shows the general procedure of a numerical flow calculation. Practical tips and empirical values can be looked up here, which are helpful for the user of CFD programs to be able to carry out CFD calculations efficiently and to be able to achieve accurate results.

Following this chapter, you should be able to answer the following questions:

1. What are the five steps in the process of a computational fluid dynamics calculation?
2. In which ways can the geometry be read in?
3. Which boundary types limit the computational domain?
4. Why do symmetry planes help to save computation time?
5. How far away from the geometry should the edges of the computational domain be? Why?
6. Where should the computational mesh be condensed?
7. What does mesh refinement strategy mean?
8. What is meant by a mesh independence study?
9. What is done in pre-processing?
10. Which initial solution should be used, the one generated from the boundary conditions or the most similar existing solution? Why?
11. How can the course of the flow be made visible?
12. What is validation?
13. What should be considered when using CFD programs for novel applications?
14. In what angle of attack range can the CFD program in Fig. 6.5 be used with good accuracy?
15. Why is it difficult to calculate larger replacement areas?

© The Author(s), under exclusive license to Springer Fachmedien Wiesbaden GmbH, part of Springer Nature 2022
S. Lecheler, *Computational Fluid Dynamics*,
https://doi.org/10.1007/978-3-658-38453-1_6

6.2   Overview

The typical workflow of a numerical flow calculation is shown in Fig. 6.1.

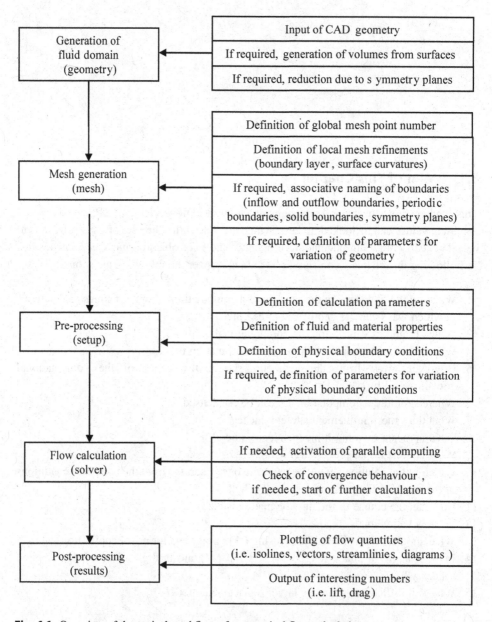

Fig. 6.1 Overview of the typical workflow of a numerical flow calculation

6.3   Generation of the Calculation Area (Geometry)

The geometry can be provided or generated in different ways:

- Usually it is provided in a 3D CAD format. Common formats are shown in the following table:

ACIS (SAT)	.sat
Parasolid	.x_t,.xmt_txt,.x_b,.xmt_bin
CATIA	.CATPart,.CATProduct
IGES	.igs,.iges
STEP	.stp,.step
solid edge	.par,.asm,.psm,.pwd
SolidWorks	.SLDPRT,.SLDASM
Autodesk	.ipt,.iam
Pro/ENGINEER	.prt,.asm
Mechanical Desktop	.dwg
OneSpace Designer	.pkg,.bdl,.ses,.sda

- The second possibility is to read in coordinates from a text or Excel file. Such coordinates usually come from design programs or from laser measurements. A polyline and a surface must then first be created from the points. The disadvantage of such models is that there are a lot of edges, which can lead to problems during mesh generation.
- For simpler geometries, these can also be created with an internal CAD program. In ANSYS WORKBENCH, this is the SPACE CLAIM program as standard from Version 17. Previously, the DESIGN MODELER program was used here, which can still be set via Tools/Options/Geometry Import/Preferred Geometry Editor. For turbomachinery there are special geometry programs with which a model can easily be created from previously determined parameters, such as BLADEGEN for pumps and turbine applications.

The computational domain is the area in whose interior the flow is to be calculated. It is limited by the geometry with the solid edge and all outer edges (see Fig. 4.1). The boundary conditions are specified later in the setup on these edges, which is why each edge must be defined separately.

For external flows such as the airfoil in Chap. 7, these are:

- The inflow edge where the flow enters the screen area and the outflow edge where the flow leaves the screen area.

- The periodic edges at which the flow periodically enters and exits the computational domain. This means that the flow that leaves the computational domain at one periodic edge re-enters the computational domain at the other periodic edge.
- Symmetry planes, at which the flow is symmetrical.
- The edges of the computational domain should be so far away from the geometry that there is a constant flow there, so that the usually constant physical boundary conditions do not distort the flow at the geometry. Usually these are three axial lengths of the geometry, e.g. three chord lengths of the aerofoil.

In the case of internal flows such as the pipe flow in Chaps. 8 and 9, only the inflow and outflow edges need to be defined as outer edges. Their position is usually already defined by the geometry. Each boundary surface can also be subdivided so that one can create different fine meshes on each subarea or define different boundary conditions such as an adiabatic wall region and a wall region with heat transfer.

To save mesh points and computation time, the computational domain should be only as large as necessary. Therefore, unnecessary components should be removed, solids should be cut away and symmetry planes should be exploited:

- Unnecessary components are, for example, attachments outside the flow. They do not influence the flow simulation and can be removed from the CAD models.
- The solids themselves can be cut away, since only the surface to the flow, i.e. the solid edge, is necessary for the flow calculation:
  - For example, in the case of the aerofoil in Chap. 7, the inside of the wing is cut out.
  - For internal flows as in Chap. 8, the pipe is cut away.
  - In the case of heat transfer as in Chap. 9, only the outer pipe is cut away. The inner pipe remains, because the heat transfer from the inner to the outer fluid through this pipe wall is to be calculated.
  - For the calculation of fluid-structure interactions (FSI), such as the thermal expansion of components or the expansion due to pressure loading, both models are necessary, the model for the structure simulation (FE model) and the model for the flow simulation (CFD model).
- Furthermore, symmetry planes with respect to the flow can be exploited:
  - The mirror symmetry (Fig. 6.2 left). Here it is possible to cut at the symmetry plane so that only one half is used, as for example with the T-piece.

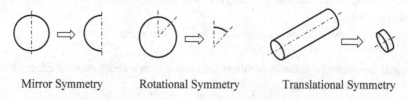

Mirror Symmetry   Rotational Symmetry   Translational Symmetry

Fig. 6.2 Examples of symmetry planes

- In the rotational symmetry in Fig. 6.2 middle, only one pie slice is meshed and computed.
- The translational symmetry (Fig. 6.2 right) can be exploited for very long components. Provided that the ends are not considered, a thin disk can be used for the simulation, such as for the infinite aerofoil. However, at least one solid element must be present; a purely two-dimensional geometry is usually not possible for computational reasons.

6.4   Generation of the Mesh (Meshing)

Modern programs for the generation of computational meshes such as MESHING, ICEM-CFD, TURBOGRID, or TGRID are very powerful and generate high-quality meshes with reasonable effort even for complex geometries. Since the accuracy of the numerical solution and the convergence behavior depend crucially on the quality and fineness of the computational mesh, the generation of a good computational mesh is very important. It usually also requires the most work for the CFD user. The following generally valid rules should therefore be observed:

- The mesh cells should be as rectangular and square as possible and the rates of change of the length of a mesh cell to the neighbouring cell should not be greater than 1.2. Then the truncation error is the smallest and the accuracy the greatest.
- The computational mesh must be compacted in areas with strong gradients. Strong gradients occur at bends and strong curvatures, as well as in the wall boundary layer and at compaction joints. Experience shows that the wall boundary layers should be resolved with at least ten points.
- Structured computational meshes that can be bisected once or twice are helpful for rapid convergence. First one computes on the coarse mesh. The high-frequency disturbances are not resolved here and are thus quickly attenuated. Then one interpolates on the medium mesh and computes some iterations before interpolating on the fine mesh to get the final solution with good resolution. This mesh refinement strategy allows stationary solutions to be obtained more quickly.

Grid generation normally takes place in several stages:

- Step 1: Check the CAD topology. To ensure that the computational domain has no errors such as holes or overlapping surfaces, the CAD topology should be checked first. Most programs offer an option for this. If the check shows errors, it is usually not possible to create a mesh.

- Step 2: Generation of the standard mesh. For this purpose, most mesh generation programs use default values that are automatically calculated from the dimensions of the computational domain. However, these standard meshes are usually too coarse, which is why they have to be manually refined globally and locally.
- Step 3: Global mesh refinement. By changing the input parameters, a sufficiently fine computational mesh can be generated. The correct parameters are usually obtained by trial and error or by using empirical values from previous calculations.
- Step 4: Local mesh refinement. The surface mesh is locally refined in the boundary layer and at strong curvatures.

For new applications, a mesh independence study should be carried out first. For this purpose, calculations are carried out on meshs of different resolution and the solutions are compared with each other. Only when the solution no longer changes does one have a sufficiently fine mesh available:

- Figure 6.3 shows schematically that only the finest mesh with $300 \times 50 \times 50$ points provides a lift coefficient that no longer changes noticeably.
- For the coarser meshes with $100 \times 20 \times 20$ and $150 \times 25 \times 25$ points, the differences are still too large and the solution thus too imprecise.
- For industrial applications, the mesh with $200 \times 35 \times 35$ points would be sufficient, since the deviation from the finest mesh is still acceptable, but the computation time is shorter. It thus represents a good compromise between accuracy and effort.

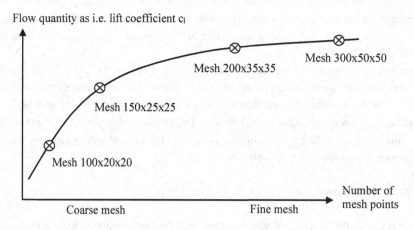

Fig. 6.3 Schematic of a mesh impact study

6.5 Preparation of the Flow Calculation (Setup)

In the so-called pre-processing or setup, all calculation parameters necessary for the solution are entered, such as:

- the type of flow (steady or unsteady)
- the material properties (fluid or solid, material values)
- the turbulence model
- the time step
- and the boundary conditions.

The convergence criteria are also defined here:

- The termination criterion of the calculation at a certain iteration number or a certain termination precision.
- Additional quantities, such as the static pressure at a certain mesh point or the lift coefficient of a wing profile, can also be defined as further convergence quantities. These quantities are displayed during the flow calculation at each calculation step.

The definition of the boundary conditions for the boundary surfaces requires some preliminary considerations from the user. The input data must be selected in such a way that they are physically meaningful in any case. The correct mass flows, pressures and temperatures must be calculated beforehand at the respective boundary surfaces. If, for example, only half a model with a symmetry plane is used, the mass flow must also be halved.

6.6 Flow Calculation (Solution)

For the flow calculation, the file generated by the pre-processing is read in together with the computational mesh.

Furthermore, a starting solution is necessary. This is usually generated by the CFD program itself from the boundary conditions during the first calculation. For further calculations it makes sense to use the "most similar" existing solution as a starting solution in order to save calculation time.

For extensive computing tasks, several processors can be used to shorten the computing time. The distribution to the processors is done automatically for most programs.

6.7   Evaluation (Results)

Numerous evaluation options are available for so-called post-processing:

- A visualization of the flow in the form of particles that swim with the flow, streamlines and flow vectors. This can also be displayed in the form of movies.
- Representation of flow variables on surfaces (isosurfaces or contours) such as pressure or temperature distributions on the wall.
- Representation of flow variables along lines in diagrams, such as the pressure curve on the wing surface.
- Tables of flow quantities, also as export for other programs.
- One-dimensional parameters such as the force on the body in the *x-direction*.
- An automatically generated report about the performed calculation with all important input data and the generated images and diagrams.

If a calculation is performed using symmetries, the rest of the model can be shown again to better illustrate the results.

All views can be saved in a graphic format such as a JPG file.

A lot of experience is needed for the correct interpretation of the results. Ultimately, the aim is to optimize the geometry with a view to increasing efficiency and reducing flow losses. At the same time, the structural-mechanical (such as durability, service life, freedom from resonance) and design constraints (such as efficient manufacturability, costs) must be observed.

6.8   Validation

Validation is the process of checking the accuracy of numerical calculation results. How well do they agree with measurements, other calculations or theory. Validation is very important when the flow is to be calculated for the first time with a CFD program for new applications.

Figures 6.4 and 6.5 show such a validation using the example of the airfoil segment from Chap. 7. Figure 6.4 shows the comparison of the measured and calculated pressure distribution at a Mach number of $Ma = 0.11$ and an angle of attack of $\alpha = 8°$ (in the low-speed wind tunnel this was the maximum possible Mach number). The agreement is very good, with only a few measurement points available in the leading edge region.

Figure 6.5 shows the lift coefficient measured in the wind tunnel and calculated with CFX. The calculated lift coefficient agrees $\alpha \leq 12°$ quite well with the measurement up to an angle of attack of. In this range the program could be used for design and optimization. For larger angles of attack, $\alpha > 12°$ the calculation provides a lift coefficient that is too large. This is due to the fact that the flow on the suction side detaches less strongly in the

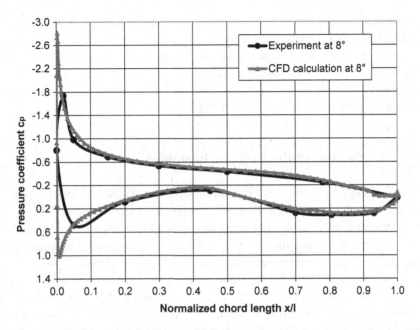

Fig. 6.4 Comparison of the measured and calculated pressure distribution on an airfoil segment at a Mach number of $Ma = 0.11$ and an angle of attack of $\alpha = 8°$

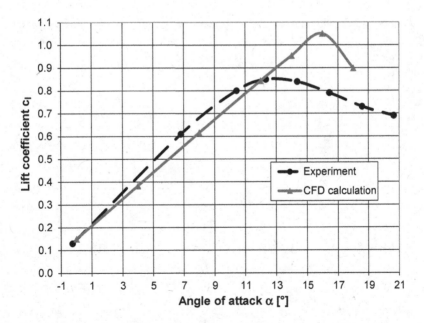

Fig. 6.5 Comparison of the measured and calculated lift coefficient on an airfoil segment at a Mach number of $Ma = 0.11$

calculation than in the experiment. In this range, the calculation program should no longer be used for the design, or only with caution.

In general, the exact steady-state calculation of larger detachment areas is difficult. On the one hand, the flow becomes unsteady due to the larger vortices and can only be calculated correctly with an unsteady calculation. On the other hand, the turbulence models used as standard are still too inaccurate for larger detachment areas.

6.9   Introduction to the Exercise Examples

In Chaps. 7, 8 and 9, the procedure for the numerical flow calculation with ANSYS CFX is shown in detail using three examples. The first two examples are intended as an introduction and are deliberately simple. The third example is a bit more complex as it deals with the heat transfer from a warm fluid through a pipe wall to a cold fluid. Further tutorials are available in the ANSYS Workbench help or on the ANSYS homepage.

The first example in Chap. 7 is the airfoil section of a modern commercial aircraft near the wing tip. The airflow is calculated here in two dimensions in order to be able to compare it with results from wind tunnel measurements (see Sect. 6.8). Figure 6.6 shows the lines of constant Mach number (Iso-Mach lines) in cruise flight at an airspeed of $Ma = 0.75$ and an

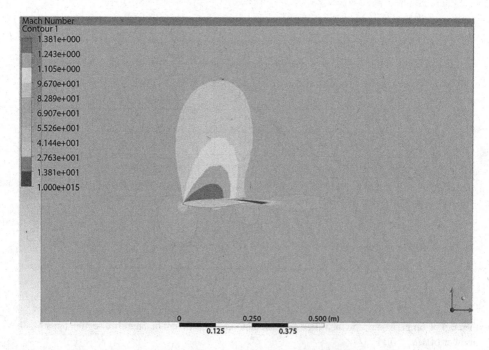

Fig. 6.6 Iso-lines of constant Mach number around an airfoil section at $\alpha = 8°$ angle of attack

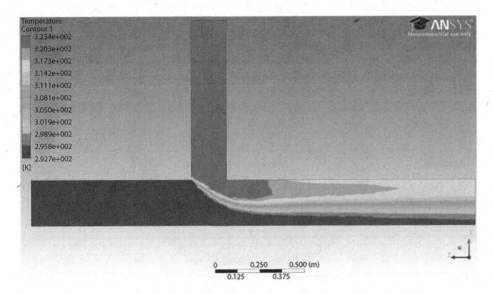

Fig. 6.7 Iso-lines of constant temperature (isotherms) during mixing in a T-pipe

angle of attack of $\alpha = 8°$. Clearly visible are the orange and red supersonic region ($Ma > 1.0$) and the blue detachment region (Ma $\approx$ 0) on the suction side.

In the second example in Chap. 8, the mixing of two water flows of different temperatures in a T-pipe section is calculated. This internal flow is three-dimensional. Figure 6.7 shows the lines of constant temperature (isotherms). Clearly visible is the penetration of the red hot water flow from above into the blue cold water flow entering from the left and the subsequent mixing until it exits on the right.

The third example in Chap. 9 deals with a typical counterflow double tube heat exchanger, where heat is transferred from a hot fluid in the inner tube through the tube wall to the cold fluid in the outer tube. Both the flow in the inner and outer tubes and the heat conduction in the intervening copper tube are calculated. Figure 6.8 shows the isotherms at the inlet of the cold water flow outside (blue) together with the outlet of the hot water flow inside (green to red) and the intervening copper pipe (light blue).

6.10   The ANSYS WORKBENCH Working Environment

For the following three exercise examples, version 18.1 was used, which was available at the time of writing the fourth edition of this book. The options and calls that must be activated or clicked are indicated in **bold**.

WORKBENCH is the common user interface for all ANSYS programs from geometry creation and mesh generation to flow simulation and strength calculation. It allows access

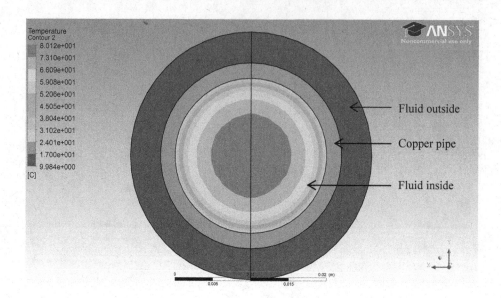

Fig. 6.8 Iso-lines of constant temperature (isotherms) for a double-tube heat exchanger

to all files created within WORKBENCH projects. Below are two important notes on error prevention:

- Entering numbers with comma or dot? The first two program parts Geometry and Mesh have a German user interface, which is why decimal numbers must be entered here with a comma. The three remaining program parts Setup, Solution and Result have an English user interface and the decimal numbers must be entered with a dot.
- Do not use umlauts like ä, ö, ü in directory and file names! This leads to a crash starting with the English-language program part Setup, since directories and files with umlauts are then no longer found.

The program is started by calling the program **ANSYS 18.1/Workbench 18.1.** either via Start/All Programs (for Windows 7) or the corresponding app (for Windows 8 and 10).

In the Workbench user interface, there is the Toolbox on the left side. For the analysis systems, click on **Fluid Dynamics (CFX) with the left** mouse button and drag the rectangle into the Project Scheme window. You can change the name to **Fluegel, for example** (Fig. 6.9). This project structure is then saved in a wbdb file when it is closed later.

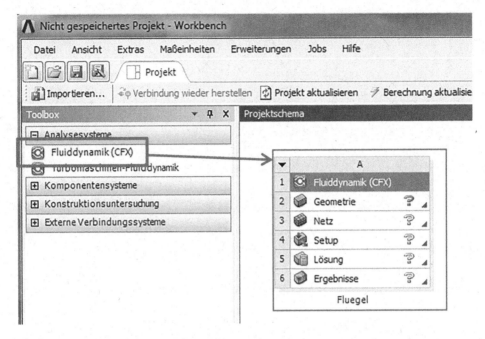

Fig. 6.9 The ANSYS WORKBENCH working environment

Example Airfoil Flow

7.1   Generation of the Calculation Area (Geometry)

- In all three exercise examples, a CAD file that was previously created separately is now read in directly. This is done as follows:
- Right-click on the **Geometry** box (see Fig. 6.9).
- Select Import Geometry
- Select the **All files (*.*)** option to display all file types.
- Select and open the directory and the CAD file (here e.g. **Fluegel.igs**).

Figure 7.1 shows the calculation area for the wing. It extends from the left inflow boundary to the right outflow boundary. At the top and bottom are the two periodic boundaries and at the front and back two symmetry planes. In the middle the wing surface is present as a solid boundary. The inside of the profile has been cut away.

The imported geometry file includes the fluid volume flowing through the airfoil. Here, the flow around a two-dimensional wing section is to be calculated. Since ANSYS-CFX requires a (three-dimensional) volume, an arbitrary span width of only 10 mm is selected. This allows the number of mesh points in the span direction to be kept low.

After the CAD file has been successfully imported, a green tick appears in the structure tree under Geometry (see Fig. 7.2).

The creation of the computational domain in the DESIGN MODELER program is no longer described here, as in the previous book editions. The reason is on the one hand that the standard geometry program from version ANSYS17 is now SPACE CLAIM and on the other hand that the geometry is usually created in a separate CAD program anyway. If

© The Author(s), under exclusive license to Springer Fachmedien Wiesbaden GmbH,   127
part of Springer Nature 2022
S. Lecheler, *Computational Fluid Dynamics*,
https://doi.org/10.1007/978-3-658-38453-1_7

0,000   1,000   2,000 (m)
0,500   1,500

Fig. 7.1 Wing: Calculation area

Fig. 7.2 Wing: Project scheme
after successful completion of
geometry import

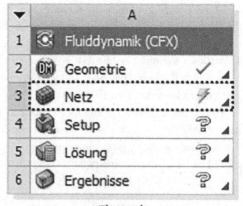

Fluegel

required, the old geometry program DESIGN MODELER can still be activated as an
option via **Tools/Options/Geometry Import/Preferred Geometry Editor: Design
Modeler**.

7.2   Generation of the Mesh (Meshing)

7.2.1   Starting the MESHING Program and Creating a Standard Mesh

To generate the computational mesh, double-click the **Mesh** field in the WORKBENCH
project schema. The MESHING program starts.

Clicking **Create mesh** in the task bar creates an initial unstructured mesh with the
default parameters provided by the program (Fig. 7.3).

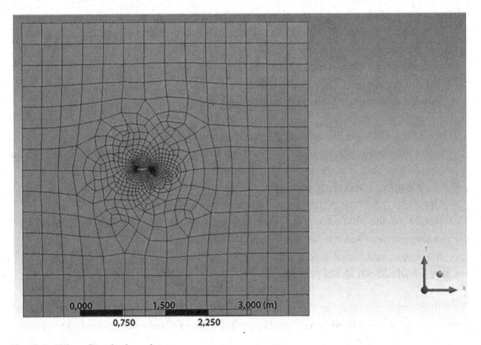

Fig. 7.3 Wing: Standard mesh

7.2.2   Refinement of the Computational Mesh on the Profile

With the default settings of the program MESHING, the flow boundary layer at the airfoil is
not well resolved and no accurate flow solution can be expected. Therefore, the computational mesh must be refined at the airfoil. This is done in the program MESHING by the
introduction of a so-called prism layer at the airfoil, a local structured O-grid of hexahedra.

In addition, with hexahedra the truncation error is smaller than with tetrahedra due to the smaller obliquity, which benefits the accuracy of the numerical solution.

The procedure is as follows:

- In the structure tree, click on **mesh** with the right mouse button, then
- Insert and
- Generation of the prism layers (inflation).

The parameters can now be entered in the detail view (in order to access the correct detail view, it may be necessary to select **Creation of Prism Layers in** the Structure Tree).

- Specify in the detailed view under area:
  - Selection method: Geometry selection
  - Geometry: 1 body
  If necessary, **click on the body/element in the** task bar (green cube).
    In the graphics window, **select the solid,** which in this case is the entire computational domain, since there is only one solid.
    Click **Apply**, 1 body appears.
- Specify in the detailed view under Definition:
  - Boundary reference Method: Geometry selection
  - Limitation: 1 surface
  If necessary, **click on the area** in the task bar (cube with 1 green area).
    In the graphics window, select the **surface** around which the prismatic layer is to be placed. In this case it is the inside of the airfoil.
    Click **Apply**, 1 surface appears.
  - Option to create the prism layers: **Thickness of the first layer**
  - Height of the first layer: **0.5 mm** (If m is set in the task bar under Units of measurement, switch to mm or enter the value in m).
  - Maximum number of layers: **10**.
- Click **Create Mesh in the** taskbar.

The refined computational mesh is generated. In the task bar, **Show mesh** must be selected if necessary so that the mesh is visible. Figure 7.4 shows the prism layer around the airfoil. The complete refined mesh is shown in Fig. 7.5.

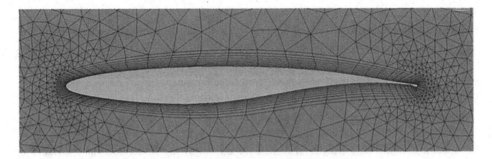

Fig. 7.4 Wing: Refined computational mesh with structured O-mesh on the profile

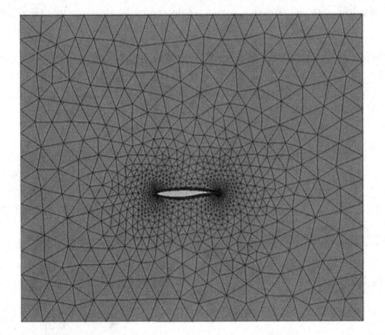

Fig. 7.5 Wing: Refined computational mesh

7.2.3 Associative Naming of the Boundaries

All surfaces have numbers as names by default. In order to find the correct surfaces for the definition of the physical boundary conditions later in the setup, it is advantageous if each boundary surface is defined separately as a component with an associative name. The procedure is as follows:

- **Mark area in the** task bar (cube with green area).
- **Mark** an **boundary area in** the graphics window, e.g. the inflow boundary on the left. This area then turns green. If necessary, the geometry must be rotated in the graphics window until this boundary area becomes visible and can be clicked on.
- Select **Create Component** with the right mouse button.
- The component has the name Selection by default. Click on it with the right mouse button in the structure tree and **rename it, e.**g. to **Inlet**.

Components should be created and named for the following seven boundary surfaces in this example: **Inflow_**, **Downflow**, **Profile**, **Periodic_up**, **Periodic_down**, **Symmetry_front** and **Symmetry_back**. These boundary surfaces are displayed in the graphics window with red arrows if they are selected in the structure tree (Fig. 7.6).

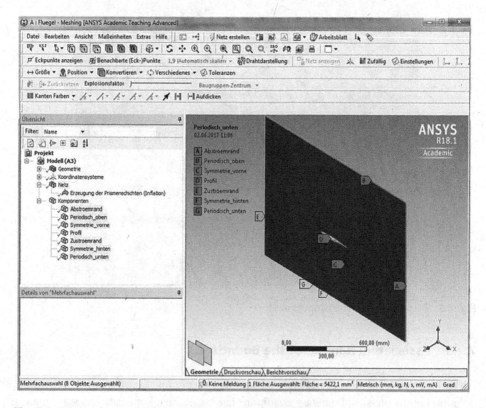

Fig. 7.6 Wing: After naming the boundary surfaces (components)

7.2.4   Exiting the MESHING Program

By closing the MESHING program in the upper right corner, the mesh data are automatically saved in a file CFX.cmdb. It is then available for the subsequent preparation of the calculation. In the WORKBENCH project scheme, a green tick now appears next to Mesh.

7.3   Preparation of the Flow Calculation (Setup)

Attention: from here on decimal numbers have to be entered with a dot according to the English notation.

7.3.1   Starting the CFX-PRE Program

To generate the input data for the flow calculation, double-click the **Setup** field in the WORKBENCH project scheme. The program CFX-PRE starts (Fig. 7.7). The computational mesh can be displayed by selecting the mesh file **CFX.cmdb** in the structure tree.

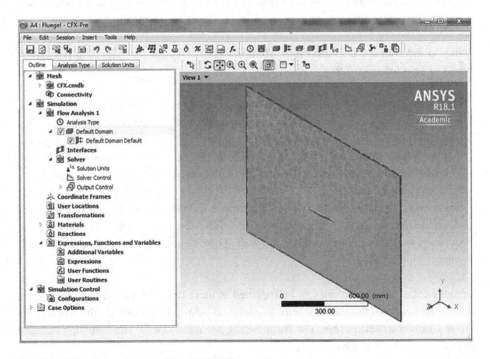

Fig. 7.7 Wing: Start window of the CFX-PRE program

7.3.2   Definition of the Calculation Parameters

Most of the calculation parameters are already set with sensible default values. In the following some parameters are given, which are of general importance or should be changed. There are always three options for closing the input windows:

- If the window can be closed with **OK, the** changes become effective and the window closes.
- If you close with **Apply, the** changes take effect and the window remains open.
- If you close the window with **Close**, the window closes without saving any changes.

In the structure tree under **Analysis Type, you** can specify whether a steady state or transient flow is to be calculated. The default value is **Steady State** and is also used for this example.

The following parameters should be defined in the structure tree under **Default Domain/ Basic Settings** (Fig. 7.8):

- Domain Type: Fluid Domain for a flow calculation (default)
- Fluid Material: **Air Ideal Gas** for a bill with ideal gas behavior.
- Reference Pressure: Here the pressure level for the calculation is specified. All pressures entered later are relative values and refer to this reference value. Entering **0 [Pa** or **bar]** has the advantage that absolute and relative pressures are identical

Switch to the **Default Domain/Fluid Models** window, where the following parameters should be selected (Fig. 7.9):

- Heat Transfer Option: **Total Energy**. In the energy conservation equation, the total energy is taken into account, which consists of internal, kinetic and potential energy (only for liquids). The Viscous Work Term option is not enabled by default. This means that the friction terms are not taken into account in the energy conservation equation. This usually affects the accuracy only slightly, but saves calculation time and improves the convergence behavior.
- Turbulence Option: **Shear Stress Transport** is the turbulence model in CFX recommended for industrial applications and exercises. It solves the k-ε *equation* and switches to the k-ω *equation near the* wall. This results in good accuracy with short computation times.

If required, the calculation run can be specified in more detail in the structure tree if you want to deviate from the standard settings specified below. To do this, double-click in the **Solver Control structure tree. The Basic Settings** window opens. The following standard parameters can be found here (Fig. 7.10):

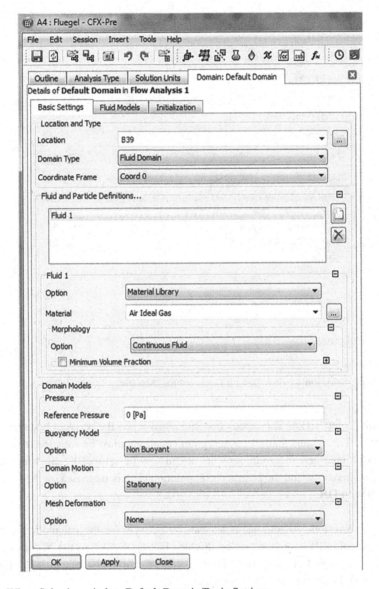

Fig. 7.8 Wing: Selection window Default Domain/Basic Settings

- Advection Scheme Option: High Resolution. This is the solution algorithm that provides good accuracy and robustness, even on shocks.
- Max. If the residuals do not fall below the target value, the calculation run stops after this number of iterations at the latest.
- Timescale Control: Auto Timescale. The time step can be entered here. It is calculated automatically for stationary calculations.

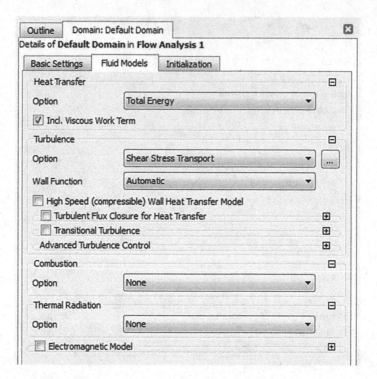

Fig. 7.9 Wing: Selection window Default Domain/Fluid Models

- Residual Target: 1.E-4. The calculation run stops when the residual becomes smaller than 10^{-4}. This means that all conservation equations must be satisfied to an error of 10^{-4}. For higher accuracy requirements, it can also be set to 10^{-5}, in which case the number of iterations and the required computation time is increased.
- Confirm with **OK**.

Furthermore, additional output variables can be defined for the calculation run. Open **Output Control** in the structure tree by double-clicking. Switch to the **Monitor** window (Fig. 7.11). Additional convergence variables can be defined here, which are displayed during the calculation run for convergence control.

- **Check** the **Monitor Objects** box. Six options appear.
- Under the **Monitor Points and Expressions option,** a net point and a flow variable can now be defined.
- Click on the **Add New Item** field to the right. The **Monitor Point 1** window appears, confirm with **OK**. A monitor point is then created whose position and variable can now be defined:
  - Option: **Cartesian Coordinates**

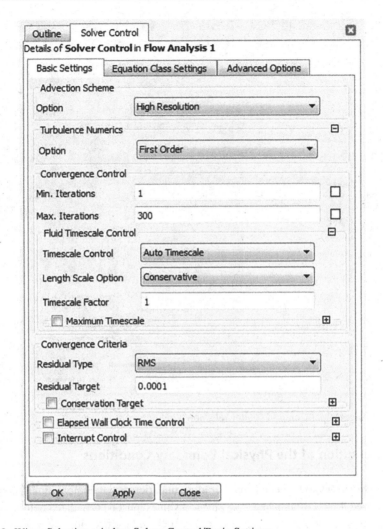

Fig. 7.10 Wing: Selection window Solver Control/Basic Settings

- Output Variables List: **Absolute Pressure**
- Cartesian Coordinates: **0 0 0**. Here the coordinates of the monitor point can be specified, such as the leading edge. Alternatively, the point can also be marked in the graphics window. It then appears as a yellow cross.
- Confirm with **OK**.

In the later flow calculation, the course of the static pressure at the leading edge of the airfoil is now displayed as a convergence course. Experience shows that the flow in the leading edge area converges the slowest, since strong gradients prevail there and the correct position of the stagnation point must first be established.

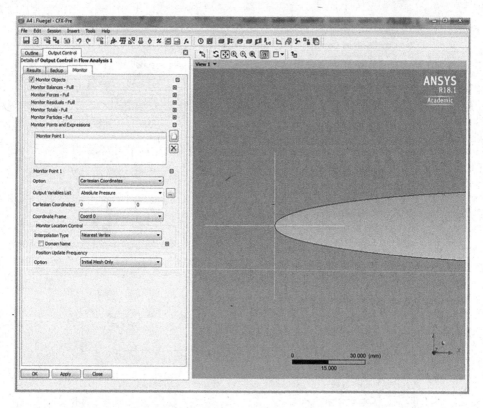

Fig. 7.11 Wing: Output Control/Monitor selection window

7.3.3   Definition of the Physical Boundary Conditions

Boundary conditions must be specified at all boundaries of the computational domain so
that the desired flow is established. The physical values must be calculated beforehand. For
the airfoil, for example, a cruise speed of $Ma = 0.75$ and a cruise altitude of $h = 11.2$ km
are specified. Based on the standard atmosphere, a static pressure of $p = 22{,}625$ Pa and a
static temperature of $T = 216.5$ K result for this flight altitude. From this, for a compressible
air flow with $\kappa = 1.4$, the total pressure and the total temperature result in

$$p_t = p \cdot \left(1 + \frac{\kappa - 1}{2} \cdot Ma^2\right)^{\frac{\kappa}{\kappa - 1}} = 32.858 \text{Pa}, \qquad (7.1)$$

$$T_t = T \cdot \left(1 + \frac{\kappa - 1}{2} \cdot Ma^2\right) = 241 \text{K}. \qquad (7.2)$$

And the overall speed to

$$v_{ges} = Ma \cdot a = Ma \cdot \sqrt{\kappa \cdot R \cdot T} = 221\frac{m}{s}. \qquad (7.3)$$

From this, the Cartesian velocity components u, v can be calculated depending on the angle of attack α of the aerofoil (for this purely two-dimensional flow, $w = 0$):

α	0°	4°	8°	12°	Grade
$u = v_{ges} \cdot \cos(\alpha)$	221.0	220.5	219.1	216.4	m/s
$v = v_{ges} \cdot \sin(\alpha)$	0.0	15.4	30.8	46.9	m/s

The physical boundary conditions can be generated in two ways:

- either via the task bar with the **Boundary** symbol (symbol with two arrows pointing to the left)
- or via the structure tree. To do this, select **Default Domain with the** right mouse button. The boundary condition is created via **Insert** and **Boundary**.

The window Insert Boundary appears, where a meaningful name for the boundary condition can be entered. This concerns the following boundaries: inflow boundary left, outflow boundary right, solid boundary at the airfoil, both symmetry planes front and back and the two periodic boundaries top and bottom.

7.3.4   Inflow Boundary

For the inflow boundary, e.g. **Inlet is** entered in the Insert Boundary window. The two new windows Basic Settings and Boundary Details appear (Fig. 7.12):

- In the **Basic Settings** window, the border type is defined and a region is assigned:
  - Boundary Type: **Inlet**
  - Location: **Zustroemrand**
- In the **Boundary Details** window, the physical boundary conditions are entered:
  - Flow Regime: **Subsonic**
  - Mass and Momentum: **Total Pressure (Stable)**
  - Relative Pressure: **32858 Pa**
  - Flow Direction: **Cartesian Components**
  - X Component: **219.1 m/s** (for $\alpha = 8°$)
  - Y Component: **30.8 m/s** (for $\alpha = 8°$)
  - Z Component: **0 m/s**
  - Turbulence: **Medium (Intensity $= 5\%$)**
  - Heat Transfer: **Total Temperature**
  - Total Temperature: **241 K**.
- Click **OK** to save the entered values.

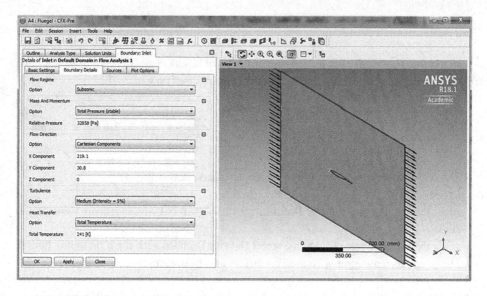

Fig. 7.12 Wing: Selection window Boundary:Inlet/Boundary Details

7.3.5 Outflow Boundary

For the outflow boundary, e.g. **Outlet is** entered in the Insert Boundary window. The two windows Basic Settings and Boundary Details appear again:

- The input in the **Basic Settings** window looks as follows:
  - Boundary Type: **Outlet**
  - Location: **Outlet**.
- In the **Boundary Details** window, a physical boundary condition is specified for a subsonic flow:
  - Flow Regime: **Subsonic**
  - Mass and Momentum: **Static Pressure**
  - Relative Pressure: **22625 Pa**.
- Click **OK** to save the entered values.

7.3.6 Solid Boundary

For the solid boundary, enter e.g. **Solid in** the Insert Boundary window.

- In the **Basic Settings** window, select:

- – Boundary Type: **Wall**
- – Location: **Profile**.
- • In the **Boundary Details** window, select:
  - – Wall Influence on Flow: No Slip. This corresponds to a frictional flow on the wall. (Free Slip corresponds to frictionless flow on the wall).
  - – Heat Transfer: Adiabatic. This corresponds to a wall without heat transfer.
- • Click **OK** to save the entered values.

7.3.7 Symmetry Planes

- • For the definition of the two symmetry planes, e.g. **Symmetry is** entered in the Insert Boundary window:
- • Select in the **Basic Settings** window:
  - – Boundary Type: **Symmetry**.
  - – Location: **Symmetry_back, Symmetry_front. Click the** three **points on the right** and **select both symmetry planes with the Ctrl key**.
- • The Boundary Details window does not appear here because no physical boundary conditions need to be specified at symmetry planes.
- • Click **OK** to save the entered values.

7.3.8 Periodic Boundary Conditions

The periodic boundary conditions are not formally treated as boundary conditions by CFX-PRE, but as so-called domain interfaces. To do this, click on the **Create a Domain Interface** icon in the task bar (two blue areas with connecting lines). The **Insert Domain Interface** window appears, where the name can be specified, e.g. **Periodicity.**

- • In the **Domain Interface** window, select:
  - – Interface Type: **Fluid Fluid**
  - – Interface Side1 Region List: **Periodic_Up** for the upper periodic border
  - – Interface Side2 Region List: **Periodic_down** for the lower periodic border
  - – Interface Models Option: **Translational Periodicity**.
- • Click **OK** to save the entered values.

If all margins are defined, the term Default Domain Default disappears in the structure tree under Outline. In the graphics window, the boundary conditions for the computational domain are indicated as arrows (Fig. 7.13).

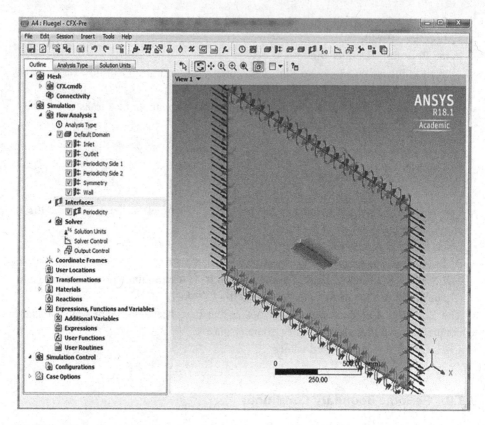

Fig. 7.13 Wing: Setup with all entries

7.3.9   Exiting the CFX-PRE Program

The data of the preparation of the flow calculation are saved in the files Fluegel.cfx and Fluegel.def when CFX-PRE is closed. After closing the program, a green tick for the setup appears in the WORKBENCH window Project scheme.

7.4   Calculation of the Flow (Solution)

7.4.1   Starting the CFX-SOLVER Program

To start the flow calculation, double-click the **Solution** field in the WORKBENCH project scheme. The CFX-SOLVER MANAGER starts.

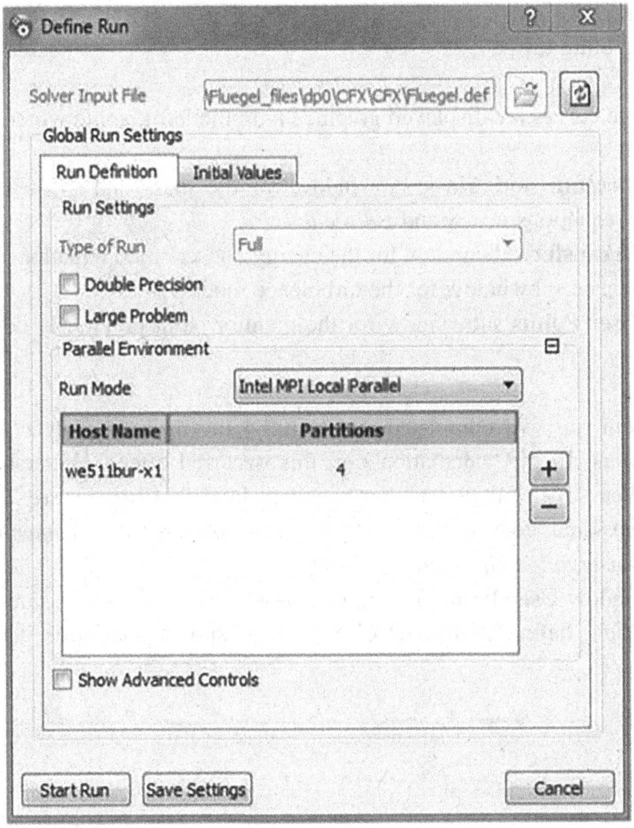

Fig. 7.14 Wing: Start window for the calculation run with CFX-SOLVER

In the window Define Run (Fig. 7.14) that opens, you can specify under **Run Mode** whether the flow calculation is to be carried out on one (Serial) or on several computers in parallel (e.g. Intel MPI Local Parallel). This allows the calculation time to be shortened.

Start the flow calculation with **Start Run.**

7.4.2 Monitoring Convergence Behaviour

The convergence behavior during the flow calculation is displayed in the CFX-SOLVER MANAGER window. Several graphic windows are visible on the left and a text field on the right. The most important variables of the calculation run are displayed in the right text window:

- the selected input variables
- the residuals for each conservation equation at each iteration step

- Warnings and error messages
- and the computing time.

The convergence curves are displayed graphically in the left graphic window:

- in the **Momentum and Mass** subwindow for the mass and the three momentum conservation equations in x, y and z *direction*
- in the **Heat Transfer** subwindow for the energy conservation equation
- in the Turbulence subwindow for the turbulence models
- and in the **User Points** subwindow for the monitor point previously defined in CFX-PRE.

If the calculation run was completed successfully, the message **Solver Run Finished Normally** appears. In this calculation run, this occurred after 233 iterations, since all residuals are then smaller than 10^{-4}, as specified in CFX-PRE. Figure 7.15 shows the courses of the residuals for the mass (red), *x-momentum* (green), *y-momentum* (blue) and *z-momentum conservation equations* (yellow).

In the subwindow **User Points the** monitor point for the static pressure at the leading edge of the airfoil, defined before in CFX-PRE, is shown above the iteration number

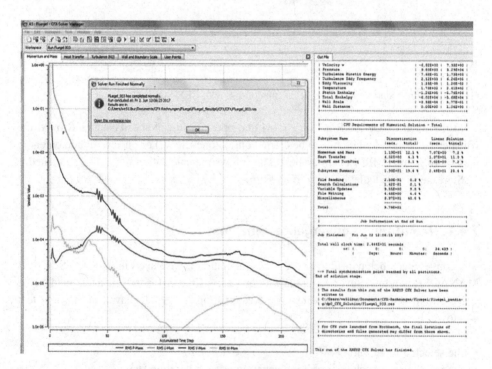

Fig. 7.15 Wing: Convergence curves after completion of the calculation run

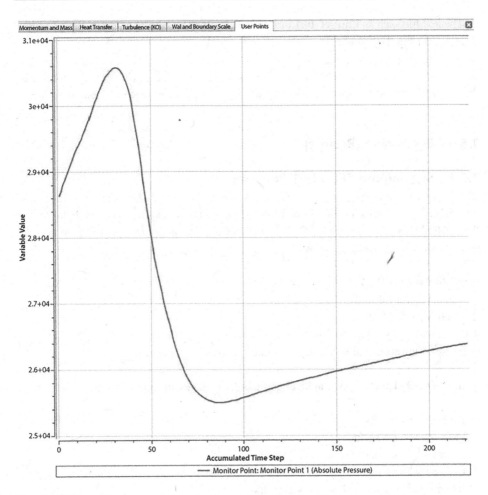

Fig. 7.16 Wing: Convergence curve for the monitor point

(Fig. 7.16). The calculation is convergent when this value no longer changes appreciably. This is approximately fulfilled here, since the *y-axis* for the pressure is very finely resolved.

7.4.3   Exiting the CFX-SOLVER Program

When closing the CFX-SOLVER program, two result files are automatically saved:

- The result file such as **Fluegel_001.res**. It contains the result and the convergence curves. The result is used for subsequent runs and the evaluation in CFD-POST. The convergence curves can be displayed again with CFX-SOLVER.

- The text file such as **Fluegel_001.out**. It contains the text displayed in the right window above during the calculation.

The number _001 is the number of the calculation run. In a subsequent run, the counter of the result files is incremented by one to Fluegel_002.res and Fluegel_002.out.

7.5   Evaluation (Results)

7.5.1   Starting the CFD-POST Program

To display the results of the flow calculation, double-click the **Results** field in the WORKBENCH project scheme. The CFD-POST evaluation program starts. Five tabs are visible below the graphics window:

- 3D Viewer for 3D views
- Table Viewer for tables
- Chart Viewer for diagrams
- Comment Viewer for comments
- Report Viewer for automatically generated reports.

With CFD-POST numerous graphical and tabular evaluations are possible:

- Isolines on surfaces
- Flow vectors on surfaces
- Streamlines and particle paths on surfaces
- Gradients along lines
- Integral values such as buoyancy and drag
- Automatically generated reports on the performed invoice.

There are also numerous options available for changing the image section. They can be accessed via the lower part of the task bar, which is located directly above the graphics window, such as:

- Rotate (two rotating arrows)
- Move (four arrows in all four directions)
- Zoom in or out (magnifying glass + and −). Alternatively, zooming in and out of the image section can also be performed using the zoom wheel of the mouse.

7.5.2   Generation of Isoline Images

For this two-dimensional example, the representation on the symmetry plane is particularly useful. For this purpose, the computational domain can be rotated so that only this plane is visible:

- **Click on** the **Z axis** in the graphics window and the *xy plane* is displayed.

To create **contour** images, click the **Contour** icon (square with colored circles) in the task bar:

- A name can be specified in the Insert Contour window. The default name is Contour1.
- In the detail view, the plane and the variable of the isolines can now be specified. e.g. for the Mach number on the symmetry plane (Fig. 7.17):
  - Locations: **Symmetry**
  - Variable: **Mach Number**
  - # of Contours: 11 is the default number of isolines.
- Click **Apply**. Contur1 then appears in the structure tree at the top left.

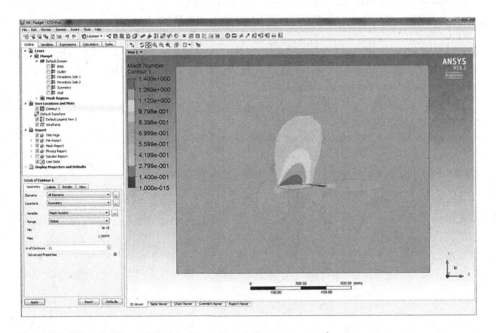

Fig. 7.17 Wing: Isolines of the Mach number on the symmetry plane

Numerous variables are available, such as pressures, temperatures and velocities. 16 variables can be selected by **clicking on the downward pointing triangle**, numerous others by clicking on the **three dots** to **the right**.

The images can also be saved. Click in the task bar to do this:

- File and Save Picture.
- In the new Print window, the file name and the output format, e.g. **Machzahl.jpeg**, can be specified.

7.5.3   Vector Image Creation

To create vector images, click the **Vector** icon (three arrows) in the task bar.

- A name can be specified in the Insert Vector window. The default name is Vector1.
- The **Contour1 isoline image** should be **deactivated in** the structure tree. Otherwise isolines and vectors are displayed superimposed in the graphics window.
- Enter in the detail view:
  - Locations: **Symmetry**
- Click **Apply** and Vector1 appears in the structure tree at the top left.

A vector is set on each node whose local length corresponds to the velocity. The global vector length is calculated automatically based on the mesh fineness. However, it can also be set manually for reasons of clarity:

- Switch to the **Symbol** tab in the detailed view
- Symbol Size: **0.2**. the global vector length is reduced to 20%.

Figure 7.18 shows the velocity vectors around the airfoil section.

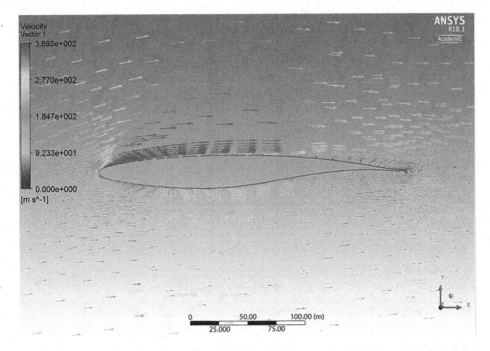

Fig. 7.18 Wing: Velocity vectors on the symmetry plane

7.5.4 Streamline Image Generation

To create streamline images, click the **Streamline** icon in the taskbar:

- A name can be specified in the Insert Streamline window. The default name is Streamline1.
- In the structure tree, the **isoline and vector images** should again be **deactivated** so that several images are not displayed on top of each other.
- Enter in the detail view:
  - Start From: **Inflow**
  - # of Points: 25. 25 streamlines are set by default.
- Click **Apply** and Streamline1 appears in the structure tree at the top left.

Figure 7.19 shows the streamlines around the airfoil section.

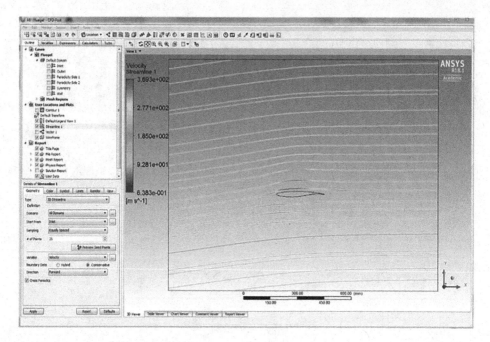

Fig. 7.19 Wing: Streamlines on the symmetry plane

7.5.5   Generation of Diagrams

To generate diagrams, the line on which the flow variable is to be displayed must first be defined. To do this, click on **Insert/Location/Polyline** in the task bar:

- A name can be specified in the Insert Polyline window. The default name is Polyline1. It then appears in the structure tree at the top left.
- Enter in the detail view:
  - Method: **Boundary Intersection**
  - Boundary List: **Wall**
  - Intersect with: **Symmetry**.
- This defines the line between the airfoil and the plane of symmetry

The second step is to define the chart. In the task bar, click the **Chart** icon (*xy-diagram* with three coloured lines).

- A name can be specified in the Insert Chart window. The default name is Chart1.
- In the detail view, the title of the diagram can be specified:
  - Title: **Pressure distribution on pressure and suction side**.
- The details are now defined in the detail view:
  - in the **Data Series** tab under Location: **Polyline 1**

- in the X Axis tab under Variable: **X**
- in the Y Axis tab under Variable: **Pressure**.
- Click **Apply** and Chart1 appears in the structure tree at the top left.

Now the pressure distribution along the pressure and suction side appears in the chart window in the Chart Viewer tab (Fig. 7.20).

These values can also be saved in a file for later processing, e.g. in Excel. For this, in the detail view:

- **Export** can be clicked.
- In the export window, the file type can be selected, such as ***.csv** or ***.txt**.
- After entering the file name, save with **Save**.

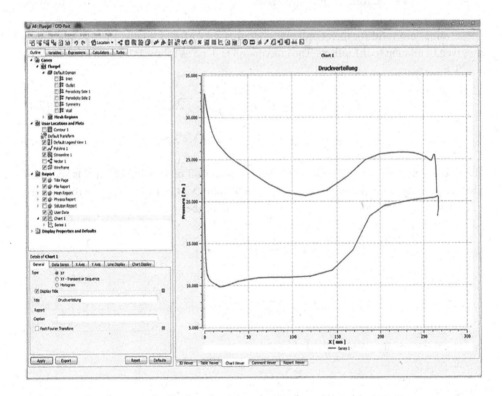

Fig. 7.20 Wing: Pressure distribution along the wing section

7.5.6 Calculation of Integral Values

In CFX-POST, integral values such as the forces on the airfoil can also be output directly. To do this, click on the Function **Calculator** symbol in the task bar (calculator with function f). Alternatively, you can switch to the **Tools** window in the structure tree and click on **Function Calculator**:

- Enter in the detail view:
  - Function **Force**
  - Location: **Wall**
  - Direction: **Global X** or **Y**.
- After clicking **Calculate,** the result appears in the Results field:
  - -0.05 [N] for the force in F_x *x-direction*
  - 25.6 [N] for the force in F_y *y-direction*.

These values are valid for the coordinate system of the airfoil. Usually lift and drag forces are given in relation to the inflow. The conversion is done as follows (for an inflow angle of $\alpha = 8°$):

$$F_a = F_y \cdot \cos (\alpha) - F_x \cdot \sin (\alpha) = 25{,}4\text{N}, \tag{7.4}$$

$$F_w = F_y \cdot \sin (\alpha) + F_x \cdot \cos (\alpha) = 3{,}5\text{N}. \tag{7.5}$$

These forces are valid for the (arbitrary) span of the airfoil of 10 mm, which was specified when generating the computational domain in DESIGN MODELER. Only the dimensionless lift and drag coefficients c_a, c_w are independent of the span width:

$$c_a = \frac{F_a}{\frac{1}{2} \cdot \rho_\infty \cdot v_{ges,\infty}^2 \cdot A}, \tag{7.6}$$

$$c_w = \frac{F_w}{\frac{1}{2} \cdot \rho_\infty \cdot v_{ges,\infty}^2 \cdot A}, \tag{7.7}$$

since by the wing area A was divided again.

7.5.7 Preparation of a Report

A report is automatically created to document the invoice. It is visible in the graphics window in the **Report Viewer** tab. Chapters of the report can also be deactivated in the structure tree under **Report.** With **Refresh Preview** in the detail view the report is then updated.

7.5.8   Exiting the CFD-POST Program

All input data created in the structure tree are automatically saved in a file Fluegel.cst when the CFD-POST program is closed. This file is reused for further evaluations when CFD-POST is called up again, whereby the same image types (but with the new results) are automatically generated again.

Example Internal Pipe Flow

8

8.1 Generation of the Calculation Area (Geometry)

After starting **ANSYS WORKBENCH** and creating the **Fluid Dynamics (CFX)** project scheme (see also Sect. 6.10), the **geometry is** started again (Fig. 8.1). In addition to importing the CAD file in Sect. 8.1.1, the generation of the geometry with the DESIGN MODELER program is described in Sect. 8.1.2.

8.1.1 Importing a CAD File

- In this exercise, too, a CAD file that was previously created separately is read in. This is done as follows:
- Click the **Geometry** box (see Fig. 8.1) with the right mouse button.
- Select Import Geometry
- Select the **All files (*.*)** option to display all file types.
- Select and open the directory and the CAD file (here e.g. **TRohr.igs**).

The imported geometry file includes the fluid volume flowing through the inside of the pipe. The inlet and outlet openings must be closed beforehand in a CAD program so that the interior becomes a volume element. To save calculation time, the symmetry is exploited and only half a pipe is used. Figure 8.2 shows the calculation area for the half tube.

After successfully importing the CAD file, a green tick appears in the structure tree under Geometry.

© The Author(s), under exclusive license to Springer Fachmedien Wiesbaden GmbH, part of Springer Nature 2022
S. Lecheler, *Computational Fluid Dynamics*,
https://doi.org/10.1007/978-3-658-38453-1_8

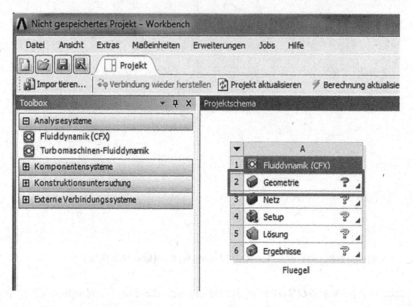

Fig. 8.1 Geometry in ANSYS WORKBENCH

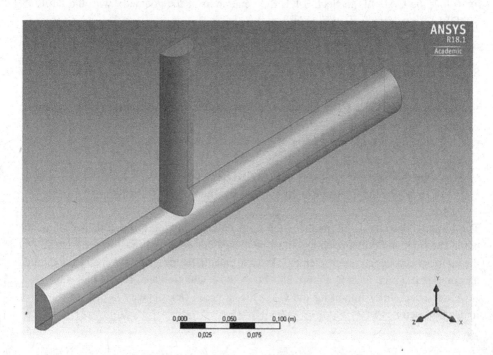

Fig. 8.2 TRohr: Calculation area

8.1.2 Generation of the Computational Domain with the Program DESIGN MODELER

Simpler geometries like this pipe branch can also be created directly in ANSYS WORK-BENCH. As of version 17, the SPACE CLAIM program is implemented here by default. Previously, the DESIGN MODELER program was used here, which can still be set via **Tools/Options/Geometry Import/Preferred Geometry Editor** (see also Sect. 7.1). For the following example, the **unit in** the top bar has been changed from **meters** to **millimeters** so that no such small numerical values have to be entered.

Start the Design Modeler program by double-clicking **Geometry** in the project scheme. Then the two cylinders are created. For the second cylinder, the specifications in () apply:

- To do this, select **Create/Basic Elements/Cylinders** in the taskbar at the top. A cylinder is created with the default name Cylinder1 (Cylinder2).
- In the detail view, the position and dimensions of the cylinders are defined:

Cylinders	Cylinder1	(Cylinder2)
Origin x-coordinate	0 mm	(0 mm)
Origin y-coordinate	0 mm	(0 mm)
Origin z-coordinate	−300 mm	(0 mm)
Axis x-component	0 mm	(0 mm)
Axis y component	0 mm	(200 mm)
Axis z component	500 mm	(0 mm)
Radius	25 mm	(20 mm)

- Then click on **Create in the** task bar.

Both cylinders now appear in the graphics window (Fig. 8.3).

In order to save calculation time during the flow calculation, symmetry planes can be used (see Sect. 6.3). In the case of the present pipe flow, one half can therefore be cut away. The procedure is as follows:

- Click **Extras/Symmetry** in the upper task bar. The default name Symmetry1 is created in the structure tree.
- In the left detail window, the plane is defined for Symmetry1:
  - Symmetry: Symmetry1
  - Symmetry plane: Select **YZ_plane** in the structure tree and click **Apply**.
- Click **Create** in the task bar at the top.

The bisected computational space can be seen in the graphics window in Fig. 8.4.

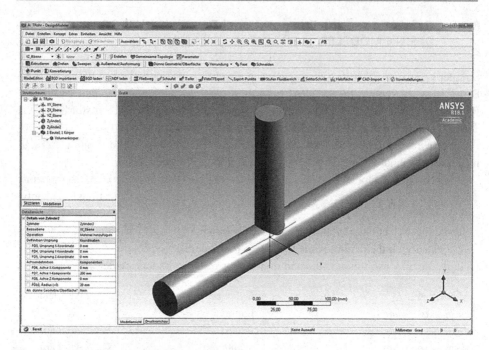

Fig. 8.3 TRohr: After creating the two cylinders

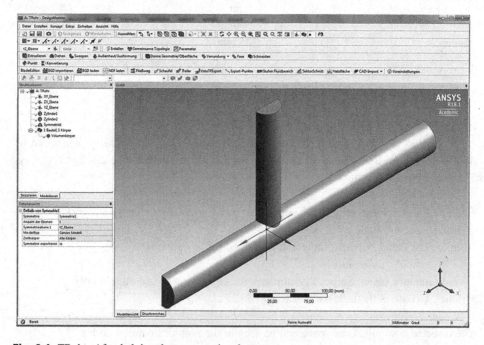

Fig. 8.4 TRohr: After halving the computational space

This geometry file is automatically saved as an **agdb file** when the DESIGN MODELER program is closed. After closing the program, a green tick appears for the geometry in the WORKBENCH window Project Scheme.

8.2   Generation of the Mesh (Meshing)

8.2.1   Starting the MESHING Program and Creating a Standard Mesh

The mesh is again created with the MESHING program. To do this, double-click on the **mesh in** the ANSYS WORKBENCH project scheme.

By clicking **Update** or **Create mesh** in the task bar, an unstructured computational mesh is created with the default parameters (Fig. 8.5). If the mesh is invisible in the graphics window, it can be activated by selecting Mesh in the Structure Tree or **Show Mesh in the** taskbar.

The standard resolution with an angle of the normal of curvature of 18° leads to the fact that a circumferential angle of 180° for half the pipe results in only 10 mesh cells in the circumferential direction. This is too coarse, which is why the entire mesh is still globally refined:

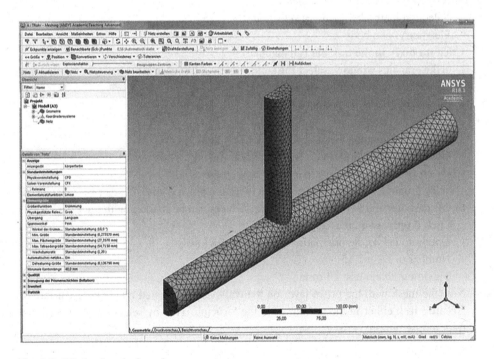

Fig. 8.5   TRohr: Standard mesh

- Select the **mesh in** the structure tree with the right mouse button.
- In the detail view, **expand the element size** (+ click):
  - Angle of the normal of curvature: **6°.**
- Click **Create Mesh in** the task bar.

Now there are 30 net cells over half the circumference of the pipe.

8.2.2   Refinement of the Mesh on the Pipe Walls

To increase the accuracy in the boundary layer, a prismatic layer (inflation) is inserted on the tube walls:

- To do this, select **Net** in the structure tree with the right mouse button, then
- Insert and
- Select generation of a prism layer (inflation).

Now the parameters of the prism layer are defined in the detail view (in the structure tree, select **Creation of** prism **layers** if necessary):

- Specify in the detailed view under area:
  - Selection method: Geometry selection
  - Geometry: one body

  Click on body in the task bar (green cube).
  In the graphics window, **select the solid**, which in this case is the entire computational domain, since there is only one solid.
  Click Apply, one body appears.
- Specify in the detailed view under Definition:
  - Boundary reference Method: Geometry selection
  - Geometry: Two surfaces

  Click on the area in the task bar (cube with green area).
  In the graphics window, **select the surfaces** on which the prism layer is to be placed.
  In this case it is the two pipe walls. Both surfaces can be marked with the CTRL key.
  Click Apply, two areas appear.
  - Option to create the prismatic layer: **Total thickness**
  - Number of layers: **15**
  - Maximum thickness: **10 mm**
- Now the mesh with the prism layer on the walls can be created by clicking **Update** or **Create Mesh** either in the taskbar or in the Structure Tree by selecting Mesh with the right mouse button.

Figure 8.6 shows the rake mesh with the prismatic layer on both pipe walls.

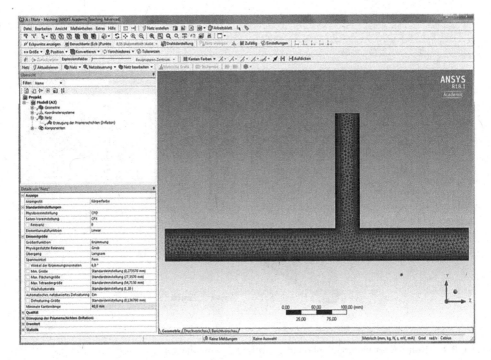

Fig. 8.6 TRohr: Refined computational mesh

8.2.3   Associative Naming of the Boundaries

In order to be able to define the physical boundary conditions correctly later in the CFX-PRE program, each boundary surface must again be defined separately as a component. The procedure is as follows:

- **Mark area in the** task bar (cube with green area).
- **Mark** an **boundary area in** the graphics window, e.g. the inflow boundary on the left. This will then turn green. If necessary, the geometry must be rotated in the graphics window until this boundary surface becomes visible and can be clicked on.
- Right-click **Insert** and select **Component.**
- The component has the name Selection by default. Click on it with the right mouse button in the structure tree and **rename it**, e.g. to **Zustroemrand_vorne**.

Components must be defined for the following five boundary surfaces in this example: **Border_front**, **Border_top**, **Border_down**, **Symmetry** and **Wall**. Multiple surfaces can be selected by holding the Ctrl key. The boundary surfaces are displayed in the graphics window with red arrows (Fig. 8.7).

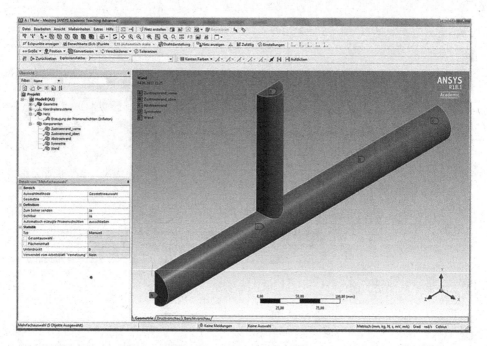

Fig. 8.7 TRohr: After naming the boundary surfaces (components)

8.2.4   Exiting the MESHING Program

This mesh file is automatically saved as CFX.cmdb when the MESHING program is closed. This file is then available for the preparation of the calculation. After closing the program, a green tick appears for the net in the WORKBENCH window Project Scheme.

8.3   Preparation of the Flow Calculation (Setup)

Attention: from here on decimal numbers have to be entered with a dot according to the English notation.

8.3.1   Starting the CFX-PRE Program

To generate the input data for the flow calculation, double-click the **Setup** field in the WORKBENCH project scheme. The CFX-PRE program starts.

The computational mesh can be displayed by selecting the **CFX.cmdb** mesh file in the structure tree.

8.3.2 Definition of the Calculation Parameters

The calculation parameters are already set with sensible default values. Only those that should be changed are specified. In the structure tree under **Default Domain/Basic Settings** these are:

- Material: **Water**
- Reference Pressure: **0 Pa**. Absolute Pressures Are Calculated. (at 1 Atm, Relative Pressures to the Reference Pressure of 1 Atm Would Be Calculated)

Switch to the **Default Domain/Fluid Models** window and change the following parameters:

- Heat Transfer Option: **Total Energy**
- Turbulence Option: Shear Stress Transport.
- Save with **OK**.

8.3.3 Definition of the Physical Boundary Conditions

Physical boundary conditions must be specified at the boundaries of the computational domain. In this example, the following boundary conditions are used:

- Mass flows are specified at the inflow boundary. It should be noted that only half the mass flow may be specified for the halved pipe.
- The averaged static pressure is specified at the downstream boundary. This has the advantage that the static pressure itself does not have to be constant at the downstream boundary. This means that the downstream area can be shortened without influencing the flow.
- A heat transfer coefficient and the ambient temperature are specified at the pipe wall.

In general, it must always be checked beforehand whether the quantities are physically meaningful and have been correctly converted!

The boundary conditions can again be generated in two ways:

- either via the task bar with the **Boundary** symbol (symbol with two arrows pointing to the left)
- or via the structure tree. To do this, select **Default Domain with the** right mouse button. The boundary condition is created via **Insert** and **Boundary**.

8.3.4   Inflow Rim in Front

- For the inflow boundary at the front, e.g. **Inlet1 is** entered in the Insert Boundary window. The two new windows Basic Settings and Boundary Details appear:
- In the **Basic Settings** window, select
  - Boundary Type: **Inlet**
  - Location: **Zustroemrand_front**.
- In the **Boundary Details** window, select:
  - Flow Regime Option: Subsonic
  - Mass and Momentum Option: **Mass Flow Rate**
  - Mass Flow Rate: **5 kg/s** (for the half pipe)
  - Flow Direction: Normal to Boundary Condition
  - Heat Transfer Option: **Static Temperature**
  - Static Temperature: **20 °C**.
- Save with **OK**.

8.3.5   Inflow Rim Top

- For example, **Inlet2 is** entered in the Insert Boundary window for the inflow boundary at the top:
- In the **Basic Settings** window, select
  - Boundary Type: **Inlet**
  - Location: **Zustroemrand_oben**.
- In the **Boundary Details** window, select:
  - Flow Regime Option: Subsonic
  - Mass and Momentum Option: **Mass Flow Rate**
  - Mass Flow Rate: **3 kg/s** (for the half pipe)
  - Flow Direction Option: Normal to Boundary Condition
  - Heat Transfer Option: **Static** Temperature
  - Static Temperature: **50 °C**.
- Save with **OK**.

8.3.6   Outflow Boundary

- For the outflow boundary, e.g. **Outlet is** entered in the Insert Boundary window:
- In the **Basic Settings** window, select
  - Boundary Type: **Outlet**
  - Location: **Abstroemrand**.
- In the **Boundary Details** window, select:
  - Flow Regime Option: Subsonic

- Mass and Momentum Option: **Average Static Pressure**
- Relative pressure: **2 bar**
- Pressure Profile Blend: 0.05
- Pressure Averaging Option: Averaging over whole outlet.
- Save with **OK.**

8.3.7 Solid Boundary

- For example, **Wall is** entered for the solid boundary in the Insert Boundary window:
- In the **Basic Settings** window, select
  - Boundary Type: **Wall**
  - Location: **Wall.**
- In the **Boundary Details** window, select:
  - Mass and Momentum Option: No Slip Wall
  - Wall Roughness: Smooth Wall
  - Heat Transfer Option: **Heat Transfer Coefficient**
  - Heat Transfer Coefficient: **400 W/m^2/K**
  - Outside Temperature: **15 °C.**
- Save with **OK.**

8.3.8 Symmetry Plane

- For the symmetry plane, for example, **Symmetry is** entered in the Insert Boundary window:
- In the **Basic Settings** window, select
  - Boundary Type: **Symmetry**
  - Location: **Symmetry.**
- The Boundary Details window does not appear here because no physical boundary conditions need to be specified at the symmetry planes.
- Save with **OK.**

If all boundary conditions are defined, Default Domain Default disappears in the structure tree and the boundary conditions are visible with arrows in the graphics window, as shown in Fig. 8.8.

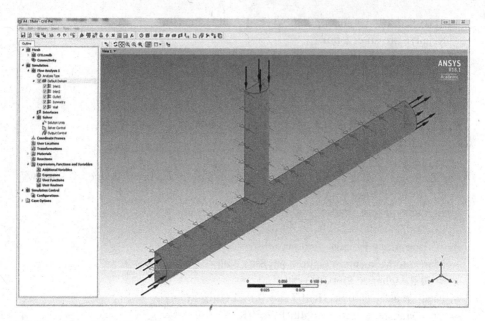

Fig. 8.8 TRohr: Setup with all inputs

8.3.9 Exiting the CFX-PRE Program

The data of the preparation of the flow calculation are saved in the files TRohr,cfx and TRohr.def when CFX-PRE is closed. After closing the program, a green tick for the setup appears in the WORKBENCH window Project scheme.

8.4 Calculation of the Flow (Solution)

8.4.1 Starting the CFX-SOLVER Program

To start the flow calculation, double-click the **Solution** field in the WORKBENCH project scheme. The CFX-SOLVER MANAGER starts.

In the Define Run window that opens, you can again specify under **Run Mode** whether the calculation run is to be executed on one (serial) or several processors (parallel). With parallel calculation the calculation time with e.g. four processors is only about a quarter of the serial calculation time on one processor.

Start the flow calculation with **Start Run.**

8.4.2 Monitoring Convergence Behaviour

After starting the calculation run, the CFX Solver tab appears. Next to the upper task bar there is again the graphics window on the left and the output list on the right.

In the left graph window, the curves of the residuals for each time step are displayed:

- The **Momentum and Mass** sub-window shows the satisfaction of the equations of mass and the three equations of conservation of momentum.
- The **Heat Transfer** subwindow shows the satisfaction of the conservation of energy equation.
- The **Turbulence** subwindow shows the fulfillment of the turbulence models.
- If monitor points have been defined in CFCXPRE, they are displayed in the **User Points** subwindow.

The right output list shows the input and output control data, warnings and error messages, the residuals at each time step, and the computation times. The calculation is terminated when either:

- The residuals of the conservation equations become smaller than the value defined in CFX-PRE (here 10^{-4}).
- The maximum number of time steps specified in CFX-PRE is reached (here 100).
- The user cancels the invoice manually. To do this, click the **stop sign in the** task bar.

In this example, the residuals of all conservation equations are smaller than 10^{-4} after 45 iterations, which automatically terminates the calculation run. Figure 8.9 shows the

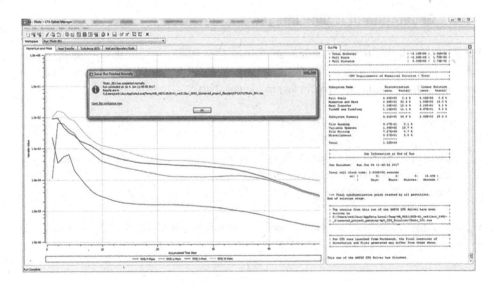

Fig. 8.9 TRohr: Convergence curves after completion of the calculation run

convergence curves for the mass conservation equation (red) and the three momentum conservation equations (x: green, y̅: blue, z: yellow).

8.4.3   Exiting the CFX-SOLVER Program

When you close the CFX-SOLVER program, two result files are automatically saved again:

- The result file such as TRohr_001.res. It contains the result and the convergence curves. The result is used for subsequent runs and the evaluation in CFD-POST. The convergence runs can be displayed again with CFX-SOLVER.
- The text file such as TRohr_001.out. It contains the text displayed in the right window above during the calculation.
- The number _001 is the number of the calculation run. In a subsequent run, the counter of the result files is increased by one to TRohr_002.res and TRohr_002.out.

8.5   Evaluation (Results)

8.5.1   Starting the CFD-POST Program

To display the results of the flow calculation, double-click the **Results** field in the WORKBENCH project scheme. The CFD-POST evaluation program starts.

8.5.2   Generation of Isoline Images

For this example, temperature isolines are generated for four surfaces, the symmetry plane, the two inflow boundaries at the front and top, and for the outflow boundary. Click on **Contour** (square with coloured circles) in the task bar:

- Enter the name. Default is Contour1
- Enter in the detail view under Geometry:
  - Locations: **Symmetry, Inlet1, Inlet2, Outlet**. The field with the three dots can be used to select several areas with the Ctrl key.
  - Variable: **Temperature**
  - # of Contours: 11.
- Click **Apply.** Contour1 appears in the structure tree at the top left.

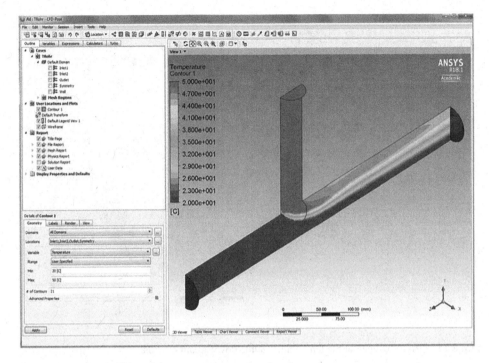

Fig. 8.10 TRohr: Isolines of the temperature on four levels

Figure 8.10 shows the isotherms at these four levels. From the left comes the colder inflow in blue and from above the warmer inflow in red. The mixing area in the right half of the pipe is clearly visible.

- The results of the half pipe can also be extended again to a complete pipe by a mirroring. To do this, click on **Contour** again in the task bar:
- Enter the name. The default is Contour2.
- Enter in the detail view under **Geometry:**
  - Locations: **Outlet**
  - Variable: **Temperature**
  - # of Contours: 11.
- Enter in the detail view under **View:**
  - Activate the checkmark in front of **Apply Reflection/Mirroring**
  - Method: *YZ-Plane*
  - X: 0.0 [m].
- Click **Apply.** Contour2 appears in the structure tree at the top left.

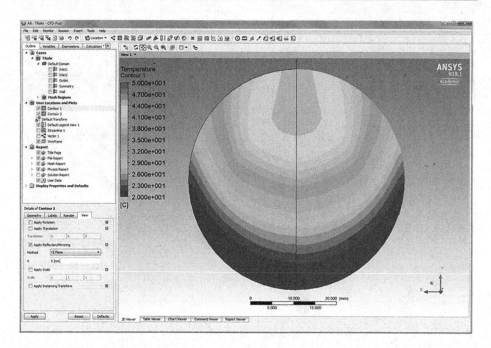

Fig. 8.11 TRohr: Isolines of the temperature at the downstream boundary

Figure 8.11 shows the isotherms at the downstream boundary over the complete pipe. The still uneven temperature distribution with 20 °C (blue) at the bottom and approx. 45 °C (orange) at the top is clearly visible.

Numerous other sizes can also be displayed on new surfaces. For this purpose, for example, a new section plane must be defined via the task bar with:

- Insert/Location/Plane.

8.5.3   Vector Image Creation

Velocity vectors are created as follows. Click on **Vector** (three arrows) in the task bar:

- Enter the name. The default is Vector1.
- Enter in the detail view under **Geometry:**
  - Locations: **Symmetry**.
- Enter in the detailed view under **Symbol:**
  - Symbol Size: **0.4**. This allows the global length of the vectors to be adjusted from the default value of 1.0 if they are too long for clear display.
- Click **Apply.**

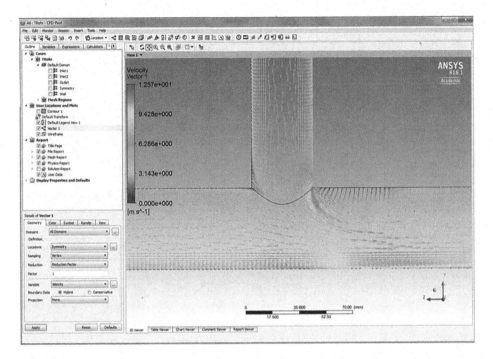

Fig. 8.12 TRohr: Velocity vectors on the symmetry plane

- In the structure tree, **deactivate the** other images such as **Contour,** otherwise the images will be displayed superimposed.
- Figure 8.12 shows the flow vectors on the symmetry plane in the area of the pipe connection. One can see here the backflow area after mixing in the upper area of the right pipe half.

8.5.4 Streamline Image Generation

Again, the streamlines are generated analogously. Click on **Streamline** in the task bar (airfoil with streamlines):

- Enter the name. The default is Streamline1.
- Enter in the detail view under Geometry:
  - Start from: **Inflow1, Inflow2**. Several surfaces can be entered by clicking on the **three dots** to the right and marking the corresponding surfaces with the **CTRL key.**
  - # of Points: 25.
- Click **Apply.**

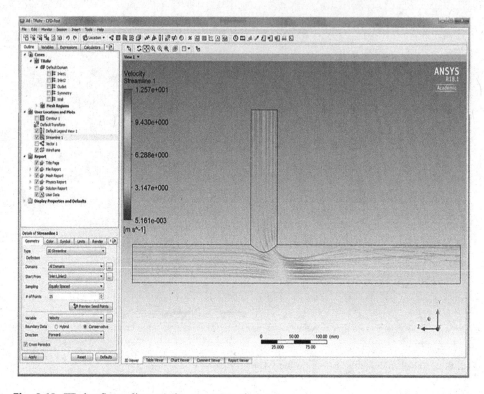

Fig. 8.13 TRohr: Streamlines on the symmetry plane

- In the structure tree, **deactivate the** other images such as **Vector** and **Contour,** otherwise the images will be displayed superimposed.

Again, in Fig. 8.13, the dead water region without streamlines is visible after mixing in the upper right pipe region.

8.5.5 Exiting the CFD-POST Program

All input data created in the structure tree are automatically saved in a TRohr.cst file when the CFD-POST program is closed. This is reused for further evaluations when CFD-POST is called up again, whereby the same image types (but with the new results) are automatically generated again.

Example Double Tube Heat Exchanger

9

9.1 Generation of the Calculation Area (Geometry)

- In this exercise, too, a CAD file that was previously created separately is read in. This is done as follows:
- Right-click on the **Geometry** box (see Fig. 6.9).
- Select Import Geometry
- Select the **All files (*.*)** option to display all file types.
- Select and open the directory and the CAD file (here **Waermerohr.igs**).

The geometry file read in comprises three parts in the case of the double-tube heat exchanger:

- the water in the inner pipe,
- the inner tube made of copper and
- the water in the outer pipe.

The outer pipe itself is not necessary, because here only the heat transfer from the inner hot fluid through the copper wall to the outer colder fluid is to be calculated.

The inlet and outlet openings must be closed beforehand in a CAD program so that the interior becomes a volume element. To save calculation time, the symmetry is exploited and only half a double tube heat exchanger is used. Figure 9.1 shows the calculation area for the half heat pipe.

© The Author(s), under exclusive license to Springer Fachmedien Wiesbaden GmbH, part of Springer Nature 2022
S. Lecheler, *Computational Fluid Dynamics*,
https://doi.org/10.1007/978-3-658-38453-1_9

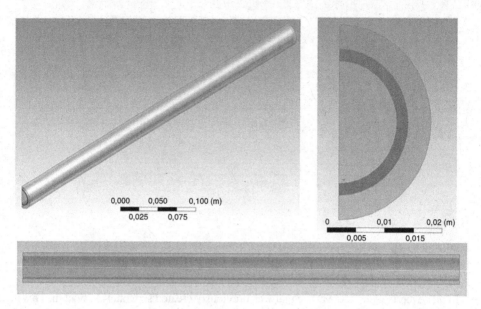

Fig. 9.1 Heat pipe: Three views of the computational domain

9.2   Generation of the Mesh (Meshing)

9.2.1   Starting the MESHING Program and Creating a Standard Mesh

The mesh is again created with the MESHING program. To do this, double-click on **Mesh** in the ANSYS WORKBENCH project scheme. The MESHING program starts and reads the previously created geometry.

Clicking **Create mesh in** the task bar creates the standard unstructured mesh. If the mesh is invisible in the graphics window, it can be activated by selecting Mesh in the structure tree.

The default resolution with an angle of curvature normal of 18° results in a circumferential angle of 180° for half the pipe resulting in only 10 mesh cells in the circumferential direction. This is too coarse, which is why the entire mesh is refined:

- Select the **mesh in** the structure tree with the right mouse button.
- In the detail view, **expand the element size** (+ click):
  - Angle of the normal of curvature: **6°.**
- Click **Create Mesh in** the task bar.

Figure 9.2 shows the refined mesh with 30 mesh cells for the half pipe.

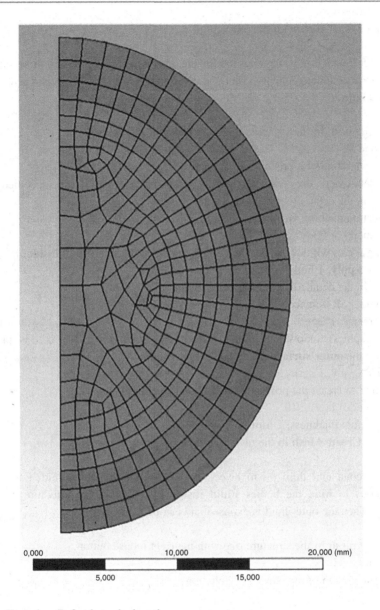

0,000   10,000   20,000 (mm)

            5,000   15,000

Fig. 9.2 Heat pipe: Refined standard mesh

9.2.2 Refinement of the Mesh on the Pipe Walls

Three prism layers are generated to increase the accuracy in the boundary layer:

* one to the outer surface of the inner fluid,
* one to the inner surface of the outer fluid and

- one to the outer surface of the outer fluid.

For the inner fluid it is advantageous to **hide** the other two bodies **pipe** and **fluid_outside** in the structure tree under geometry. Then the inner fluid is exposed. The prism layer is again created as follows:

- Select the **mesh in** the structure tree with the right mouse button.
- Insert and
- Select generation of a prism layer (inflation).
- In the detail view, the parameters for the first prism layer can now be specified:
- Specify in the detailed view under **area:**
  - Selection method: Geometry selection
  - Geometry: 1 body

  In the graphics window, select the corresponding solid, here **Fluid_inside**.
     Click Apply, 1 body appears.
- Specify in the detailed view under **Definition:**
  - Boundary Reference Method: Geometry selection
  - Geometry: 1 face

  In the graphics window, select the surface on which the prism layer is to be placed, in this case the **outer surface of** the inner fluid volume.
     Click Apply, 1 surface appears.
  - Option to create the prismatic layer: **Total thickness**
  - Number of layers: **10**
  - Maximum thickness: **2 mm**
- Click on **Create Mesh in** the task bar

Now the second and third prism layers are created for the outer fluid. For this it is advantageous to **hide** the bodies **Fluid_inside** and **pipe in** the structure tree under Geometry. Then the outer fluid is exposed and can be marked easily:

- Select the **mesh in** the structure tree with the right mouse button.
- Insert and
- Select generation of a prism layer (inflation).

In the detail view, the parameters for the second and third prism layer can now be specified:

- Specify in the detailed view under **area:**
  - Selection method: Geometry selection
  - Geometry: 1 body

  Select the corresponding solid in the graphics window, here **Fluid_outside**. **Click Apply,** 1 body appears.

- Specify in the detailed view under **Definition:**
  - Boundary Reference Method: Geometry selection
  - Geometry: 1 face

  In the graphics window, select the surface on which the prism layer is to be placed. This is the **inner surface for** the second prismatic layer and the **outer surface of** the outer fluid volume for the third prismatic layer.

  Click Apply, 1 surface appears at a time.
  - Option to create the prismatic layer: **Total thickness**
  - Number of layers: **10**
  - Maximum thickness: **1.5 mm**
- Click on **Create Mesh in** the task bar

Now all bodies have to be faded in again. Figure 9.3 shows all three prism layers.

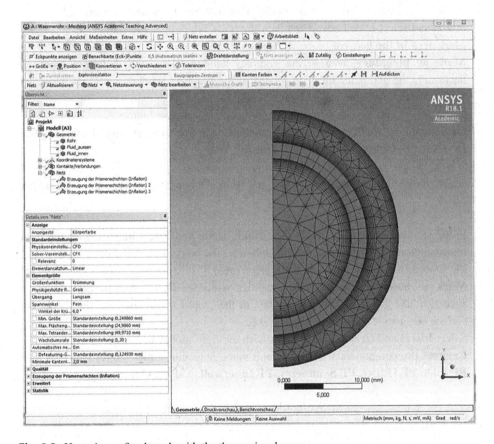

Fig. 9.3 Heat pipe: refined mesh with the three prism layers

9.2.3 Associative Naming of the Border Areas

All surfaces have a number by default. At this point it makes sense to convert the numbers into meaningful names so that they can be easily found in the CFX-PRE program when assigning boundary conditions. For this purpose, components are created again for all boundary surfaces. In order to make internal surfaces such as the wall between the fluid inside and the pipe visible, the other bodies must be hidden.

Since this example is a counterflow heat exchanger, the inflow boundaries are on opposite sides, on the left for the inner fluid (z = 500 mm) and on the right for the outer fluid (z = 0 m). This should be taken into account when naming the heat exchanger so that the correct inflow and outflow surfaces can be assigned during setup.

At inner fluid there are four border-surfaces (Inlet, Outlet, Symmetry, Wall) for which components are inserted:

- In the structure tree first mark the bodies Pipe and Fluid_outside with the right mouse button and select Hide body. Then only the body Fluid_inside is visible.
- Mark **area in the** task bar (cube with green area).
- **Mark** an **boundary area** in the graphics window. This will then turn green. If necessary, the geometry must be rotated in the graphics window until this boundary surface becomes visible and can be clicked.
- Select **Insert** and **Create Component with** the right mouse button.
- **Rename** the component with the default name Selection to a meaningful name such as **IF_Inlet, IF_Outlet, IF_Symmetry,** and **IF_Wall**.

There are five boundary surfaces for the pipe (Inlet, Outlet, Symmetry, Inner Wall, Outer Wall):

- In the structure tree, mark the bodies **Fluid_inside** and **Fluid_outside** with the right mouse button and select **Hide body.** Only the tube body is then visible.
- Mark **area in the** task bar (cube with green area).
- **Mark** an **boundary area** in the graphics window. This will then turn green. If necessary, the geometry must be rotated in the graphics window until this boundary surface becomes visible and can be clicked.
- Select **Insert** and **Create Component with** the right mouse button.
- **Rename the** component with the default name Selection a meaningful name such as **PI_Inlet, PI_Outlet, PI_Symmetry, PI_Innerwall,** and **PI_Outerwall**.

And at outer fluid also exist five border-surfaces (Inlet, Outlet, Symmetry, Inner Wall, Outer Wall):

- In the structure tree, mark the bodies **Fluid_inside** and **Pipe** with the right mouse button and select **Hide body.** Only the body Fluid_outside is then visible.

- Mark **area in the** task bar (cube with green area).
- **Mark** an **boundary area** in the graphics window. This will then turn green. If necessary, the geometry must be rotated in the graphics window until this boundary surface becomes visible and can be clicked.
- Select **Insert** and **Create Component with** the right mouse button.
- **Rename the** component with the default name Selection to a meaningful name such as **OF_Inlet, OF_Outlet, OF_Symmetry, OF_Innerwall,** and **OF_Outerwall**.

The boundary areas are indicated in the graphics window with red arrows (Fig. 9.4).

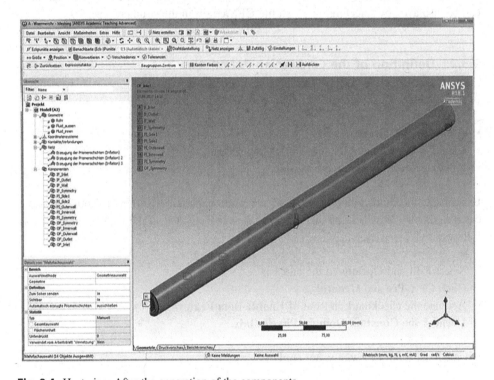

Fig. 9.4 Heat pipe: After the generation of the components

9.2.4 Exiting the MESHING Program

The mesh file is again automatically saved as **CFX.cmdb** when the MESHING program is closed. After closing the program, a green tick appears for the mesh in the WORKBENCH window Project Scheme.

9.3   Preparation of the Flow Calculation (Setup)

Attention: from here on decimal numbers with a dot have to be entered again according to the English notation.

9.3.1   Starting the CFX-PRE Program

To generate the input data for the flow calculation, double-click the **Setup** field in the WORKBENCH project scheme. The CFX-PRE program starts. The computational mesh can be displayed by selecting the mesh file **CFX.cmdb** in the structure tree.

9.3.2   Definition of the Calculation Parameters

In this example, the computational domain consists of three parts:

- A fluid area with water inside,
- a solid state region for the intervening copper tube and
- a fluid area with water on the outside.

By default, the fluid regions consist of the same material. If different fluids are to be calculated, such as water and air, this must first be activated in the CFX Pre-Options as follows:

- Select **Edit** and **Options in** the task bar
- Click **CFX-Pre** and **General**
- Activate the checkmark in front of **Enable Beta Features**
- Uncheck the **Constant Domain Physics box.**

For the inner fluid area with water applies:

- In the structure tree, select **Flow Analyis 1** with the right mouse button.
- Select **Insert/Domain** or select Insert Domain (blue box) in the task bar.
- Give it a meaningful name, such as **IF** for Inner Fluid.
- In the left structure tree, open the **IF** calculation area by double-clicking on it.
- Specify in the **Basic Settings** tab (see Fig. 9.5 left):
  - Location: IF
  - Domain Type: Fluid Domain
  - Material: **Water**
  - Reference Pressure: 1 atm

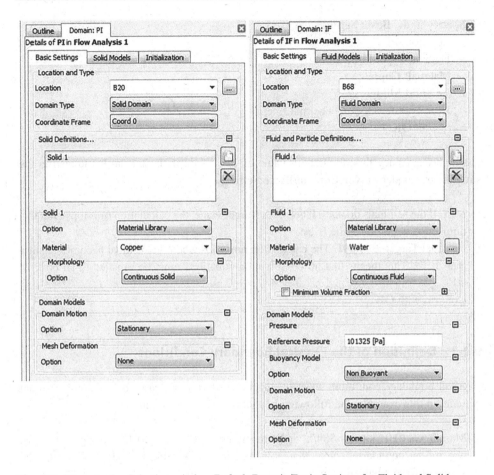

Fig. 9.5 Waermerohr: Selection window Default Domain/Basic Settings for Fluid and Solid

- In the **Fluid Models** tab:
  - Heat Transfer Option: **Total Energy**
  - Turbulence Option: **Shear Stress Transport.**
- Save with **OK**.

The same is to be repeated for the outer fluid area (here e.g. **OF**). The parameters in Basic Settings and in Fluid Models are automatically taken over from the inner Fluid IF.

For the solid state region, the values for copper are set:

- In the structure tree, select **Flow Analyis 1** with the right mouse button.
- Select **Insert/Domain** or select Insert Domain (blue box) in the task bar.
- Enter a meaningful name, e.g. **PI** for Pipe.
- In the structure tree on the left, open the calculation area **PI** by double-clicking on it.

- Specify in the **Basic Settings** tab:
  - Location: PI
  - Domain Type: **Solid Domain**
  - Material: **Copper**
- In the **Solid Models** tab:
  - Heat Transfer Option: Thermal Energy
- Save with **OK**.

If a higher accuracy than the default value of 10^{-4} is desired, this can be defined in the structure tree under **Solver Control/Basic Settings:**

- Max. If the residuals do not fall below the target value, the calculation run stops after this number of iterations at the latest.
- Residual Target: **0.00001**. The calculation run stops when the residual becomes smaller than 10^{-5}. This means that all conservation equations must be satisfied to an error of 10^{-5}.
- Confirm with **OK**.

9.3.3 Definition of the Physical Boundary Conditions

Physical boundary conditions must be specified at the boundaries of the computational domain. In this example, the following boundary conditions are used:

- Mass flows (0.01 kg/s $\cong$ 36 l/h) and temperatures (80°C and 10°C) are specified at the two inflow boundaries. Note that only half the mass flow, i.e. 0.005 kg/s, may be specified for the halved pipe.
- At the two downstream boundaries, the averaged static pressure of 0 Pa is specified as a deviation from the reference pressure of 1 atm. Averaging has the advantage that the static pressure itself does not have to be constant at the downstream boundary. This means that the downstream area can be shortened without influencing the flow.
- An adiabatic wall (no heat flow) is specified at both ends of the copper tube and for the outer wall of the external fluid.
- In the copper tube, the heat conduction is to be included in the calculation. Therefore, no boundary conditions must be specified here, but so-called interface planes must be defined.

In general, it must always be checked beforehand whether the quantities are physically meaningful and have been correctly converted!

The boundary conditions can again be generated in two ways, either via the task bar with the **Boundary** symbol (symbol with two arrows pointing to the left) or via the Structure Tree. For this, the respective **calculation area (here IF, PI, OF) is to be marked with the right mouse button**. The boundary condition is generated via **Insert** and **Boundary**.

9.3.4 Inflow Boundaries for Fluid Inside and Fluid Outside

- For the inflow boundaries of the inner or outer pipe, e.g. **IF_Inlet** or **OF_Inlet is** entered in the Insert Boundary window. The two new windows Basic Settings and Boundary Details appear.
- In the **Basic Settings** window, select
  - Boundary Type: **Inlet**
  - Location: **IF_Inlet** or **OF_Inlet**.
- In the **Boundary Details** window, select:
  - Flow Regime Option: Subsonic
  - Mass and Momentum Option: **Mass Flow Rate**
  - Mass Flow Rate: **0.005 kg/s** (for the half pipe)
  - Flow Direction: Normal to Boundary Condition
  - Heat Transfer Option: **Static Temperature**
  - Static Temperature: **80°C** for the fluid inside or **10°C** for the fluid outside.
- Save with **OK**.

9.3.5 Outlet Boundaries for Fluid Inside and Fluid Outside

- For the outflow boundary, e.g. **IF_Outlet** or **OF_Outlet is** entered in the Insert Boundary window.
- In the **Basic Settings** window, select
  - Boundary Type: **Outlet**
  - Location: **IF_Outlet** or **OF_Outlet**.
- In the **Boundary Details** window, select:
  - Flow Regime Option: Subsonic
  - Mass and Momentum Option: **Average Static Pressure**.
  - Relative pressure: **0 Pa** (referred to the reference pressure of 1 atm)
  - Pressure Profile Blend: 0.05
  - Pressure Averaging Option: Averaging over whole outlet
- Save with **OK**.

9.3.6 Solid Boundaries for Fluid Outer and Tube End Faces

- For example, **OF_Wall** and **PI_InOut** are entered in the Insert Boundary window for the solid boundary of the outer fluid region and the end faces of the pipe.
- In the **Basic Settings** window, select
  - Boundary Type: **Wall**
  - Location: **OF_Outerwall** or **PI_Inlet, PI_Outlet** (with Ctrl-key)
- In the **Boundary Details** window is already selected:

- Mass and Momentum Option: No Slip Wall
- Wall Roughness: Smooth Wall
- Heat Transfer Option: Adiabatic
- Save with **OK**.

9.3.7   Symmetry Planes for Fluid Inside, Fluid Outside and Pipe

- For the symmetry plane, enter e.g. **IF_Symmetry** or **OF_Symmetry** or **PI_Symmetry** **in** the Insert Boundary window.
- In the **Basic Settings** window, select
  - Boundary Type: **Symmetry**
  - Location:
      IF_Symmetry for the fluid inside
      OF_Symmetry for the fluid outside
      PI_Symmetry for the pipe
      The Ctrl key can be used to select multiple surfaces.

- Save with **OK**.

9.3.8   Interface Planes Between Fluid and Pipe

The connection levels between the fluid and solid computational domains are treated as domain interfaces:

- To do this, click the **Create a Domain Interface** icon (two blue areas with connecting lines) in the taskbar.
- The **Insert Domain Interface** window appears, where the name can be specified, such as **IF_PI** or **OF_PI**:
- In the **Domain Interface** window, select:
  - Interface Type: **Fluid Solid**
  - Interface Side 1 Domain: **IF** or **OF**
  - Interface Side 1 Region List: **IF_Outerwall** or **OF_Innerwall**
  - Interface Side 2 Domain: **PI**
  - Interface Side 2 Region List: **PI_Innerwall** or **PI_Outerwall**
  - Interface Models Option: General Connection.
- Click **OK** to save the entered values.

If all boundary conditions are defined, Default Domain Default disappears in the structure tree and the boundary conditions are visible with arrows in the graphics window (Fig. 9.6).

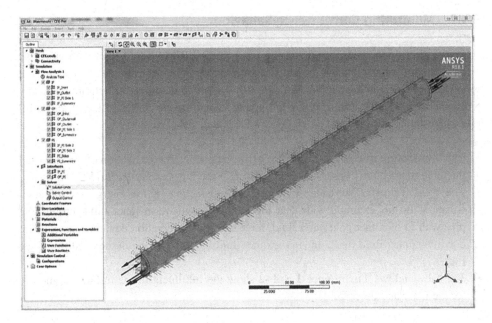

Fig. 9.6 Heat pipe: Setup with all inputs

9.3.9   Exiting the CFX-PRE Program

The data of the preparation of the flow calculation are saved in the files Waermerohr,cfx and Waermerohr.def when CFX-PRE is closed. After closing the program, a green tick for the setup appears in the WORKBENCH window Project scheme.

9.4   Calculation of the Flow (Solution)

9.4.1   Starting the CFX-Solver Program

To start the flow calculation, double-click the **Solution** field in the WORKBENCH project scheme. The CFX-SOLVER MANAGER starts. In the Define Run window that opens, start the flow calculation with **Start Run.**

The parameters specified there mean:

- Solver Input File: the input file for CFX. By default, this is the last generated setup file, such as Waermerohr.def.
- Initialization Option: Current Solution Data (if possible). For the first calculation, the program calculates the initial solution from the boundary conditions. For subsequent runs, the last solution is used.

9.4.2 Monitoring Convergence Behaviour

After the start of the calculation run the CFX-SOLVER-MANAGER appears with the graphic window on the left and the output list on the right. In this example, the residuals of all conservation equations are smaller than 10^{-4} after 72 iterations, which automatically terminates the calculation run. In the left graphic window, the curves of the residuals for each time step are displayed:

- The **Momentum and Mass-1** and **-2** subwindows show the satisfaction of the mass and the three momentum conservation equations for the inner and outer fluid.
- The **Heat Transfer** subwindow shows the satisfaction of the energy conservation equation. In this example with heat transfer from one warm inner fluid through the pipe wall to the other cold outer fluid, three energy conservation equations are solved, one in the inner fluid (red), one in the outer fluid (green) and one in the solid (blue, see Fig. 9.7).
- The sub-windows **Turbulence-1** and **-2** show the fulfilment of the turbulence models for the inner and outer fluid.

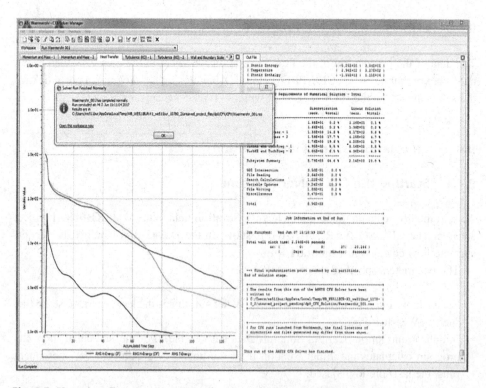

Fig. 9.7 Heat pipe: Convergence curves after the end of the calculation run

9.4.3 Exiting the CFX-Solver Program

Again, two result files are saved automatically:

- The result file such as **Waermerohr_001.res**. It contains the result and the convergence curves.
- The text file such as **Waermerohr_001.out**. It contains the text displayed in the right window above during the calculation.
- The number _001 is the number of the calculation run. In a subsequent run, the counter of the result files is increased by one to Waermerohr_002.res and Waermerohr_002.out.

9.5 Evaluation (Results)

9.5.1 Starting the CFD-POST Program

To display the results of the flow calculation, double-click the **Results** field in the WORKBENCH project scheme. The CFD-POST evaluation program starts.

9.5.2 Generation of Isoline Images

Figure 9.8 shows the isotherms on the symmetry plane and the inflow and outflow boundary. In the task bar click **Insert** and **Contour** or alternatively the square with the coloured circles:

- Enter the name. The default is Contour1.
- Enter in the detail view under Geometry:
  - Domains: **All Domains**
  - Locations: **IF_Symmetry, OF_Symmetry, PI_Symmetry** (with Ctrl key)
  - Variable: **Temperature**
  - # of Contours: 11.
- Click **Apply.** Contour1 appears in the structure tree at the top left.

The 80°C warm water flows from the left to the right in the inner pipe, transferring heat through the copper pipe to the cold water outside. Since it is supposed to correspond to a counterflow heat exchanger, the 10°C cold water in the outer pipe flows from right to left.

The isotherms at the inflow and outflow boundaries are shown in Figs. 9.9 and 9.10. For a better overview the semicircle was mirrored at the symmetry plane to get a full circle again. Click on **Insert** and **Contour in** the task bar or alternatively on the square with the coloured circles:

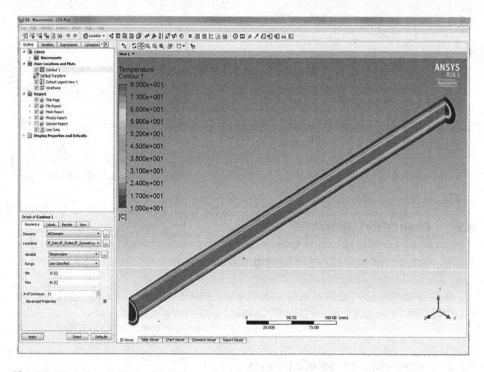

Fig. 9.8 Heat pipe: Isolines of the temperature on the symmetry plane and at the inflow and outflow boundaries

- Enter the name. The default is Contour2.
- Enter in the detail view under **Geometry:**
  - Domains: All Domains
  - Locations: **IF_Inlet, IF_Outlet, OF_Inlet, OF_Outlet, PI_InOut**
  - Variable: **Temperature**
  - # of Contours: 11.
- Click **Apply.** Contour2 appears in the structure tree at the top left.
- In the structure tree, select **Contour2** with the right mouse button and click on **Duplicate.** Contour3 is created.
- Enter in the detail view of **Contour3** under **View:**
  - Activate the checkmark in front of **Apply Reflection/Mirroring**
  - Method: *YZ-Plane*
  - X: 0.0 [m].
- Click **Apply.** Contour3 appears in the structure tree at the top left.

Numerous other sizes can also be displayed on new surfaces. For this purpose, for example, a new section plane must be defined via the task bar with **Location/Plane**.

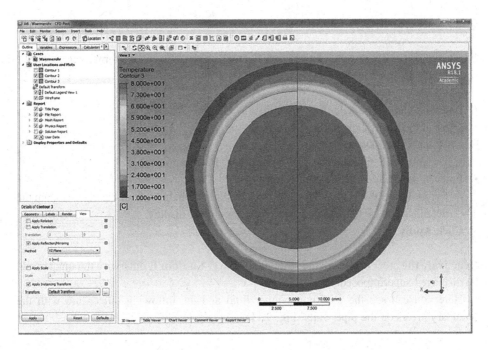

Fig. 9.9 Heat pipe: Isolines of the temperature at $z_2 = 500$ mm (warm inflow inside)

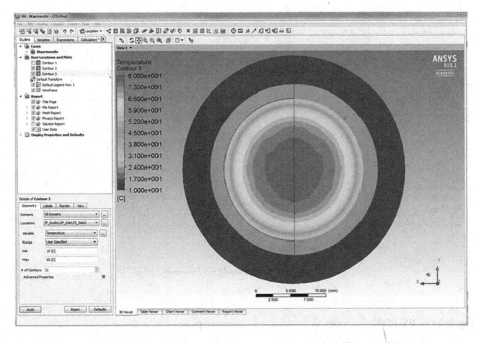

Fig. 9.10 Heat pipe: Isolines of the temperature at $z_1 = 0$ mm (cold inflow outside)

9.5.3   Vector Image Creation

For the velocity vectors, click **Insert** and **Vector in the** task bar, or alternatively click the square with the three arrows:

- Enter the name. The default is Vector1.
- Enter in the detail view under **Geometry:**
  - Domains: All Domains
  - Locations: **IF_Symmetry, OF_Symmetry, PI_Symmetry**.
- Enter in the detailed view under **Symbol:**
  - Symbol Size: **0.1**. This can be used to adjust the global length of the vectors from the default value of 1.0 if they are too long for a clear display.
- Click **Apply**.

Deactivate the other images in the structure tree, e.g. **Contour1,** otherwise the images will be displayed superimposed.

Here you can see the outflowing inner fluid and the inflowing outer fluid with the boundary layers on the pipe walls (Fig. 9.11).

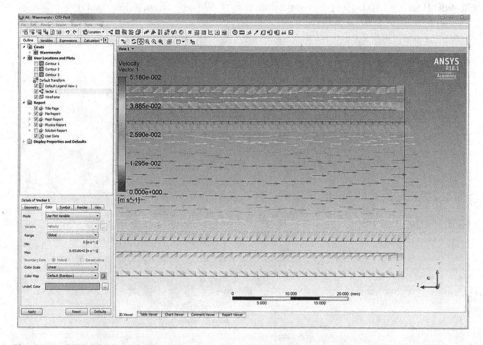

Fig. 9.11 CFX-POST: Velocity vectors on the symmetry plane at $z_1 = 0$ mm

9.5.4 Generation of Diagrams

In order to be able to display the temperature curves along the *z-axis* from $z_1 = 0$ mm to z_2 $= 500$ mm, two lines are defined at $x = 0\ mm$, the one for the warm inner fluid at the radius of $y = 0$ mm and the one for the colder outer fluid at the radius of $y = 14$ mm. To do this, click on **Location/Line** in the task bar:

- A name can be specified in the Insert Line window. The default name is Line1. It then appears in the structure tree at the top left.
- Enter in the detailed view of Line1:
  - Domains: **IF**
  - Method: **Two Points**

Point 1:	0	0	0
Point 2:	0	0	500

- Confirm with **Apply.**

Mark **Line1** in the structure tree with the right mouse button and click on **Duplicate.** Line2 is created.
   Enter in the detailed view of Line2:

- Domains: **OF**
- Method: **Two Points**

Point 1:	0	14	0
Point 2:	0	14	500

- Confirm with **Apply**.

Now the two lines along the *Z-axis are* defined for the inner and outer fluid.
   The second step is to define the chart. In the task bar, click the **Chart** icon (*xy-diagram* with three coloured lines):

- A name can be specified in the **Insert Chart** window. The default name is Chart1.
- In the detailed view of Chart1, the following data is specified:
  - Register General/Title: **Inner Fluid Temperature**
  - Register Data Series/Location: **Line1**
  - Register X Axis/Data Selection/Variable: **Z**
  - Click on **Invert Axis** so that the inflow boundary with 80°C is on the left.

- – Register Y Axis/Data Selection/Variable: **Temperature**.
- Click **Apply** and Chart1 appears in the structure tree at the top left.

In the structure tree, select **Chart 1** with the right mouse button and click on **Duplicate.** This creates Chart2.

- In the detailed view of Chart2, the data will then read:
  - – Register General/Title: **Outer Fluid Temperature**.
  - – Register Data Series/Location: **Line 2**
- Click **Apply** and Chart 2 appears in the structure tree at the top left.

Now the temperature curve along the flow for the inner fluid or the outer fluid appears in the graphic window (Figs. 9.12 and 9.13).

The warm water in the inner pipe cools down from 80°C to 77.8°C, while the cold water in the outer pipe heats up from 10°C to 19°C. The water in the inner pipe cools down from 80°C to 77.8°C, while the cold water in the outer pipe heats up from 10°C to 19°C.

These values can also be saved in a file for later processing, e.g. in Excel. For this, in the detail view:

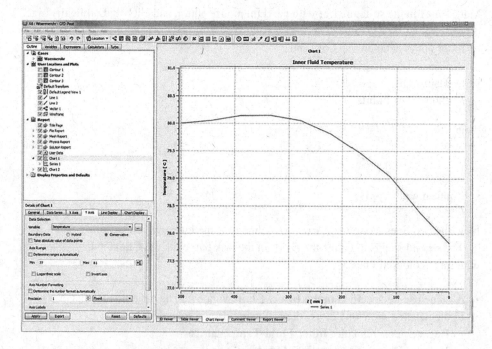

Fig. 9.12 Heat pipe: Temperature curve of the inner fluid

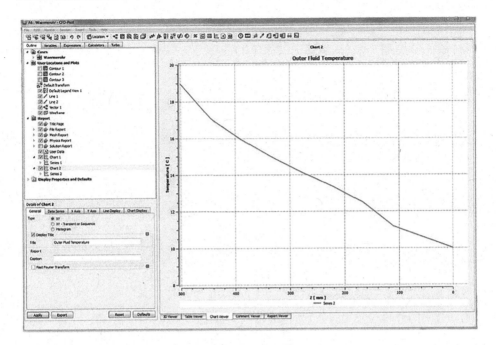

Fig. 9.13 Heat pipe: Temperature curve of the external fluid

- **File/Export can** be clicked.
- In the export window, the file type can be selected, such as ***.csv** or ***.txt**.
- After entering the file name, save with **Save**.

Furthermore, images can also be created directly via the task bar with **File** and **Save Picture**.

Example Parameter Variation

10

Often several CFD calculations are to be carried out, in which geometry or flow variables change. In the case of the aerofoil flow exercise, this can be, for example, the angle of attack or the ratio of the two velocity components u and v. *In the case of the tube flow exercise, this can be, for example, the mass flow and in the case of the double tube flow exercise, the mass flow.* In the case of the tube flow example, this could be the mass flow, and $\dot{m}$ in the case of the double tube heat exchanger example, the tube length l. Then these variables to be varied can be defined once as parameters. The values of these parameters are then defined in a parameter set. The calculation then only has to be started once, all subsequent runs with changed parameters then run automatically one after the other.

By clicking on the **Parameter Set** field in the ANSYS WORKBENCH, all previously defined parameters can be displayed (Fig. 10.1).

In the following example, three calculations were carried out for the double-tube heat exchanger, whereby the length was varied from 500 mm to 1000 mm and 1500 mm. For this purpose, the *z-axis* was defined as a parameter for all cylinders during creation in the DESIGN MODELER program by clicking the box so that a D appears. **Length** was entered as parameter name. Figure 10.2 shows the structure tree for this input parameter defined to 500 mm. To the right of the structure tree the table of design points appears (Fig. 10.3). If now calculations for other lengths are to be carried out automatically, duplicate the current design point by clicking the current design point with the right mouse button and enter the other pipe lengths.

After **closing the parameter set in** the upper task bar, the ANSYS WORKBENCH user interface appears again. If the **Update all design points** field is then clicked in the upper task bar, the required programs start automatically for each design point created:

- In the DESIGN MODELER program, the pipe length is then first adjusted.

© The Author(s), under exclusive license to Springer Fachmedien Wiesbaden GmbH, part of Springer Nature 2022
S. Lecheler, *Computational Fluid Dynamics*,
https://doi.org/10.1007/978-3-658-38453-1_10

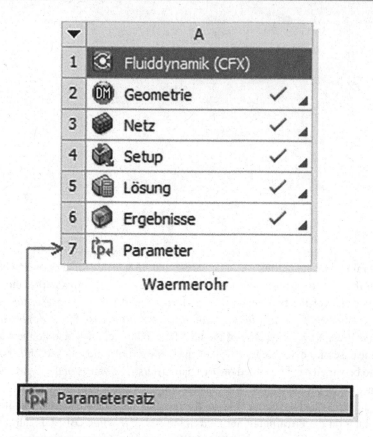

Fig. 10.1 ANSYS WORKBENCH: Project scheme with parameter set

	A	B	C	D
Strukturbaum für Alle Parameter				
1	ID	Parametername	Wert	Einheit
2	⊟ Eingabeparameter			
3	⊟ 🔘 Waermerohr (A1)			
4	⼁p P1	Laenge	500	mm ▼
*	⼁p Neuer Eingabeparameter	Neuer Name	Neuer Ausdruck	
6	⊟ Ausgabeparameter			
*	p⼁ Neuer Ausgabeparameter		Neuer Ausdruck	
8	Diagramme			

Fig. 10.2 ANSYS WORKBENCH: Structure tree for parameters

	A	B	C	D	E
1	Name ▼	Aktualisierungsreihenfolge ▼	P1 - Laenge ▼	☐ Exportiert	Hinweis ▼
2	Einheit		mm ▼		
3	Aktuell	1	500		
4	DP 1	2	1000	☐	
5	DP 2	3	1500	☐	
*				☐	

Tabelle von Design Points

Fig. 10.3 ANSYS WORKBENCH: Table with additional new design points 1000 mm and 1500 mm

- updated the computational mesh in the MESHING program
- in the program CFX-SOLVER the calculation is carried out iteratively
- and in the program CFD-POST the result representations are generated.

It can take some time until all invoices are finished.

Answers to the Target Control

Answers to Sect. 2.1 Conservation Equations of Fluid Mechanics

1. Conservation variables are the mass, the momentum in x-, y- and z-direction and the energy.
2. The differences between integral and differential forms are summarized in Table 2.1.
3. The equation of conservation of mass is derived for a volume element fixed in space. The sum of all mass flows through the six surfaces is equal to the temporal change of the mass in the volume element.
4. The momentum conservation equations are derived from Newton's second law.
5. The equation of conservation of energy is derived from the first law of thermodynamics.
6. The vector form of the Navier-Stokes equations is given as Eq. 2.18.
7. Additional equations are the thermal and the caloric equations of state and nine relations for the stresses, such as the Stokes relations.
8. Physical boundary conditions must be specified by the user, numerical boundary conditions are calculated by the program.
9. Four physical boundary conditions must be specified at the subsonic inflow boundary, none at a supersonic outflow boundary (in 3D).
10. The full Navier-Stokes equations resolve the turbulence and therefore require an extremely fine computational mesh. The Reynolds-averaged Navier-Stokes equations use turbulence models and do not require such fine computational meshes, which is why the computation times are significantly shorter.
11. Turbulence models are necessary to calculate the high-frequency turbulent fluctuations that are not resolved by the computational grid.
12. For the thin-layer Navier-Stokes equations, the friction and heat conduction terms in the two directions parallel to the wall are neglected.
13. In the Euler equations, all friction and heat conduction terms are neglected. They are therefore only valid for high Reynolds numbers and detachment-free flows.

© The Author(s), under exclusive license to Springer Fachmedien Wiesbaden GmbH, part of Springer Nature 2022
S. Lecheler, *Computational Fluid Dynamics*,
https://doi.org/10.1007/978-3-658-38453-1

14. In the potential equation, all friction and heat conduction terms are neglected as in the Euler equations, and losses are not allowed (rotational freedom or isentropic and isenthalpic flow).

15. The polar curve (lift and drag as function of angle of attack) can only be calculated with the Navier-Stokes equations, because at larger angles of attack the flow can detach.

16. To estimate the impact position around a hypersonic aircraft, the Euler equations would suffice.

Answers to Sect. 3.1 Discretization of the Conservation Equations

1. Discretization means the transformation of the differentials into finite differences.

2. Discretization methods are the finite element method, the finite volume method and the finite difference method.

3. The spatial derivatives can be discretized by forward differences, backward differences or central differences.

4. The order of accuracy is the magnitude of the truncation error. For example, for first order accuracy the truncation error is $O(\Delta x)$, for second order $O(\Delta x^2)$.

5. In the case of a continuous flow, the accuracy should be of the second order; in the case of a discontinuous flow, it should only be of the first order in order to avoid oscillations at the discontinuity point.

6. The time derivative can be discretized by a forward difference or a central difference.

7. In the time-asymptotic calculation, a stationary solution is sought and the time derivative is used only for numerical reasons. In the time-asymptotic calculation, the transient solution is sought and the time derivative is resolved physically.

8. For the transient calculation, a central difference is used because it has a second order accuracy.

9. For stationary solutions, the time derivatives are also solved, since then the conservation equations remain hyperbolic and can be solved with a solution procedure.

10. After discretization, one obtains the so-called difference equations.

11. The difference equation is consistent with the differential equation if, for mesh sizes approaching zero, the truncation errors also approach zero, i.e. the difference equations agree with the differential equations.

12. A numerical solution method is stable if the termination errors become smaller and smaller during the calculation run. The numerical solution then satisfies the difference equation.

13. A numerical solution is convergent if it satisfies the differential equations. The residual as a measure for the satisfaction of the conservation equations must always become smaller. If the residuals of all conservation equations are smaller than 10^{-4}, one usually speaks of a convergent solution.

14. An additive numerical viscosity is needed in the central spatial discretization to dampen the truncation errors, make the method stable, and reduce oscillations at discontinuity points.

15. Upwind discretization means a one-way discretization that takes into account the propagation direction of information.

16. In the explicit discretization, the flow terms are formed at the known time n, while in the implicit discretization they are formed at time $n + 1$.

17. The advantage of implicit discretization is that larger time steps can be used. As a result, the total computing times are usually shorter than with explicit methods. The disadvantage is the more complex programming, since a system of equations must be solved or a matrix must be inverted.

18. The CFL (Courant-Friedrichs-Levy) number couples the time step Δt to the mesh size Δx (see Eq. 3.30).

19. In a purely explicit procedure, the maximum CFL number can be at most 1.

Answers to Sect. 4.1 Computational Meshes

1. To have support points for the difference equations.

2. A good mesh should be as rectangular as possible with rates of change less than 1.2. Then the truncation errors are smallest and the accuracy is greatest (Sect. 6.4).

3. For skewed meshes, the mesh lines adapt to the wall contour, increasing the accuracy of the solution at the boundary.

4. O-, C-, and H-meshes are skewed meshes. The letters indicate the shape of the mesh lines (Sect. 4.3.2). O- grids resolve the boundary layer at the geometry well, but not the wake. C- and H-grids solve the boundary layer and the wake well. However, H- grids waste mesh points in the inflow.

5. Calculation meshes are compacted at the solid body surface in order to be able to resolve the flow gradients in the wall boundary layer and to increase the accuracy. With inviscid computation, this is not necessary, so that the meshes can be significantly coarser at the solid boundary, which saves computation time.

6. The boundary layer should be resolved with at least ten mesh cells in the case of a viscid calculation.

7. The advantage of block-structured meshes is that they can be composed of several blocks. This makes it possible to generate good meshes with low skewness even for more complex geometries.

8. Adaptive meshes automatically adjust their mesh size to the flow gradients. Thus, for example, shock waves can be well resolved, while unnecessary mesh points are avoided in the remaining flow field. The calculations can thus be performed accurately and efficiently.

9. Unstructured meshes are very flexible and adapt well to complex and composite geometries.

10. An ideal mesh cell should ideally be rectangular, because then the truncation errors are small and the accuracy is high.

Answers to Sect. 5.1 Solution Methods

1. The three classes of solution methods are central methods, upwind methods, and high-resolution methods.
2. Central methods are accurate, but have problems at discontinuity points such as compression shocks. The solution usually oscillates there. Upwind methods are very robust and oscillation-free, but not accurate enough. High-resolution methods combine both. They are accurate and prevent oscillations at discontinuities.
3. Among the central solution methods are the Lax-Wendroff methods, the Runge-Kutta methods and the ADI methods.
4. An advantage of the upwind methods is their robustness, even in hypersonic flows with strong compression shocks. The disadvantage is that they only have a first order accuracy and are therefore too inaccurate for steady flow.
5. Monotone solution methods prevent the occurrence of extreme points. This prevents oscillations and unphysical solutions from occurring in the first place.
6. Methods that have a spatial accuracy of second order are no longer strictly monotonic, since only first order methods satisfy this.
7. For second order methods, the TVD condition was therefore introduced, a weakened monotonicity condition. It prevents the occurrence of oscillations in methods of second order accuracy. However, it does not exclude unphysical solutions.
8. Therefore, for methods with a spatial accuracy of second order, the entropy condition must be satisfied in addition to the TVD condition. It only allows solutions in which the total entropy increases.
9. Limiter functions limit the slopes of the cell sizes at discontinuity points. This reduces the spatial accuracy at discontinuity points to first order and prevents oscillations.
10. High-resolution methods or TVD methods provide accurate and oscillation-free solutions for both steady and unsteady flow.

Answers to Sect. 6.1 Typical Workflow of a Numerical Flow Calculation

1. In a numerical flow calculation, the following five steps must be carried out: Generation of the computational domain, generation of the computational mesh, preparation of the flow calculation (pre-processing), flow calculation and evaluation (post-processing).
2. The geometry can usually be read in as a CAD file or as a text file with the coordinates. Modern CFD programs also have program parts with which the geometry can be generated directly.

3. The following boundary types limit the computational domain: inflow boundaries, outflow boundaries, periodic boundaries and symmetry planes. The fluid can flow through them. Boundaries that are not flowed through are the solid boundaries, which are formed by the geometry itself.

4. Symmetry planes help to save computation time, since the computational domain can be significantly reduced, saving mesh points and computation time.

5. Usually, the in- and outflow boundaries of the computational domain should be three characteristic geometry lengths away from the geometry itself, such as the chord length in the airfoil, so as not to distort the flow at the geometry, since constant values are often prescribed at these boundaries.

6. The mesh is to be compacted in areas with strong gradients in order to be able to achieve better resolution and accuracy. This is the case with strong curvatures and kinks, in the boundary layer and at discontinuity points such as compaction joints.

7. A mesh refinement strategy is useful to get a convergent solution faster. The calculation is started on a coarse mesh. After some iterations, the solution is interpolated to a finer mesh. On this fine mesh again some computational steps are performed before interpolating again on the next finer mesh. In this way, the higher-frequency disturbances are eliminated more quickly.

8. In a mesh independence study, calculations are performed on meshes of different fineness and their solutions are compared. The goal is to find the mesh with the smallest mesh score where the solution has almost no differences to the finest mesh.

9. During pre-processing the flow calculation is prepared and all necessary parameters and boundary conditions are defined.

10. The most similar existing solution should be used as the starting solution, because then the calculation converges the fastest.

11. The flow can best be visualized by means of velocity vectors or streamlines.

12. Validation means comparing a numerical solution with theoretical or experimental results to ensure that the computational program produces reliable results.

13. When CFD programs are used for novel applications, validation should first be performed.

14. In the range from 0° to 12°, the calculation and measurement in Fig. 6.5 agree well. In this range, the CFD program can be used with good accuracy.

15. The calculation of larger detachment areas is difficult because the flow becomes unsteady and the turbulence models are often too inaccurate.

References

1. Wendt, J.F.: Computational fluid dynamics. Springer, Berlin Heidelberg (2010)
2. Ferziger, J.H., Peric, M.: Numerische Strömungsmechanik. Springer, Berlin Heidelberg (2008)
3. Hirsch, C.: Numerical computation of internal and external flows – volume 1 fundamentals of numerical discretization – volume 2 computational methods for Inviscid and viscous flows. Elsevier, Amsterdam (2007)
4. Laurien, E., Oertel, H.: Numerische Strömungsmechanik. Springer, Berlin Heidelberg (2013)
5. Peyret, R., Taylor, T.D.: Computational methods for fluid flow. Springer, Berlin Heidelberg (1982)
6. ANSYS CFX-Solver Benutzeranleitungen, ANSYS CFX Release 18.1, 2017
7. Beam, R.M., Warming, R.F.: An implicit finite-difference algorithm for hyperbolic system in conservation law form. J. Comput. Phys. **22**, 87–109 (1976)
8. Briley, W.R., McDonald, H.: Solution of the three-dimensional Navier-Stokes equations by an implicit technique. Proceedings of 4th International Conference on Numerical Methods in Fluid Dynamics. Lecture Notes in Physics, Vol. 35. Springer, Berlin Heidelberg (1975)
9. Chima, R.V.: NASA Glenn research center, Cleveland, USA (2011). http://www.grc.nasa.gov/www/rte/images/Swift/stf1_x.jpg. Accessed on 21.10.2017
10. Courant, R., Isaacson, E., Reeves, M.: On the solution of nonlinear hyperbolic differential equations by finite differences. Comm. Pure. Appl. Math. **5**, 243–255 (1952)
11. Enquist, B., Osher, S.: Stable and entropy satisfying approximations for transonic flow calculations. Math. Comput. **34**, 45–75 (1980)
12. Godunov, S.K.: A difference scheme for numerical computation of discontinuous solution of hydrodynamic equations. Math Sbornik. **47**, 271–306 (1959) in Russian, Translation US Joint Publ. Res. Service JPRS 7226, 1969
13. Harten, A., Lax, P.D., Van Leer, B.: On upstream differencing and Godunov-type schemes for hyperbolic conservation laws. Siam Rev. **25**, 35–61 (1983)
14. Jameson, A., Schmidt, W., Turkel, E.: Numerical simulation of the Euler equations by finite volume method using Runge-Kutta time stepping schemes. AIAA Paper 81–1259 (1981)
15. Lax, P.D., Wendroff, B.: Systems of conservation laws. Comm. Pure Appl. Math. **13**, 217–237 (1960)
16. Lecheler, S.: Ein voll-implizites 3-D Euler-Verfahren zur genauen und schnellkonvergenten Strömungsberechnung in Schaufelreihen von Turbomaschinen. Fortschrittbericht VDI Reihe 7 Nr. 216. VDI-Verlag, Düsseldorf (1992)

© The Author(s), under exclusive license to Springer Fachmedien Wiesbaden GmbH, part of Springer Nature 2022
S. Lecheler, *Computational Fluid Dynamics*,
https://doi.org/10.1007/978-3-658-38453-1

17. Lerat, A.: Implicit methods of second order accuracy for the Euler equations. AIAA Paper 83–1925 (1983)
18. MacCormack, R.W.: The effect of viscosity in hypervelocity impact cratering. AIAA Paper. **69–354** (1969)
19. Osher, S., Chakravarty, S.R.: High resolution schemes and the entropy condition. Siam J. Numer. Analysis. **21**, 966–984 (1984)
20. Pulliam, T.H., Steger, J.L.: Implicit finite difference simulations of three dimensional compressible flows. AIAA J. **18**, 159–167 (1980)
21. Pulliam, T.H., Chaussee, D.S.: A diagonal form of an implicit approximate factorization algorithm. J. Comput. Phys. **39**, 347–363 (1981)
22. Roe, P.L.: The use of the Riemann problem in finite difference schemes. Lecture Notes in Physics, Bd. 141. Springer, Berlin, S. 354–359 (1981)
23. Roe, P.L.: Some contributions to the modelling of discontinuous flows. Lect. Appl. Math. **22**, 357–372 (1985)
24. Steger, J.L., Warming, R.F.: Flux vector splitting of the inviscid gas-dynamic equations with application to finite difference methods. J. Comput. Phys. **40**, 263–293 (1981)
25. Van Leer, B.: Towards the ultimate conservative difference scheme. A second order sequel to Godunov's method. J. Comput. Phys. **32**, 101–136 (1979)
26. Van Leer, B.: Flux vector splitting for the Euler equations. In: Proceedings of 8th International Conference on Numerical Methods in Fluid Dynamics. Springer, Berlin Heidelberg (1982)
27. Von Neumann, J., Richtmyer, R.D.: A method for the numerical computations of hydrodynamical shocks. J. Math. Phys. **21**(3), 232 (1950)
28. Yee, H.C.: Construction of explicit and implicit symmetric TVD schemes and their applications. J. Comput. Phys. **68**, 151–179 (1985)
29. Yee, H.C., Harten, A.: Implicit TVD schemes for hyperbolic conservation laws in curvilinear coordinates. AIAA Paper. **85–1513** (1985)

Index

A

Additive numerical viscosity, 47, 60–62, 64, 88, 99
Adhesion condition, 81
ANSYS CFX, 1, 11, 122, 127
ANSYS WORKBENCH, 115, 123, 125, 155–157, 159, 174, 195–197

B

Backward difference, 50–53, 56, 62, 77, 88, 91, 93
Baldwin Lomax model, 37, 38
Body forces, 17
Boundary, 13, 30–33, 35, 36, 38, 39, 42–44, 67–72, 78, 81–83, 90, 91, 113, 115–119, 127, 129–132, 138–141, 150, 160–165, 168, 170, 175–179, 182–185, 187, 188, 190, 191
Boundary condition
    numeric, 13, 30, 31, 36
    physical, 33, 42, 140
Boundary layer equations, 43
Boundary layer profile, 32, 81

C

CAD, 11, 68, 115–117, 127, 155, 173
Calculation parameters, 119, 134, 163, 180–182
Cell center, 48, 49, 93, 95
Cell corner, 48, 49
CFD-POST, 145, 146, 153, 168, 172, 187, 197
CFL number, 65, 88–90
CFX PRE, 133, 141, 142, 144, 161, 162, 166, 167, 178, 180, 185
CFX-SOLVER, 142–143, 145, 166, 168, 185, 197
Characteristics, 31–35, 62, 65, 90, 91, 93, 96, 106
C mesh, 71, 72

Component, 17, 22, 33, 34, 38, 42, 116, 117, 131, 132, 139, 157, 161, 162, 178, 179, 195
Compression shock, 6, 42, 44, 54, 61, 62, 85, 87, 88, 90, 91, 100, 106, 108
Compressive force, 17
Computational domain, 31, 35, 55, 61, 69, 70, 83, 113, 115–118, 127, 130, 138, 141, 147, 152, 157, 160, 163, 174, 180, 182, 184
Computational mesh
    adaptive, 10, 67, 82
    block structured, 78, 79
    Cartesian, 70, 71, 83
    oblique, 90
    structured, 81, 82, 129, 131
    unstructured, 80, 82, 159
Computational mesh adaptation, 81–83
Conservation equation, 2, 10, 13–44, 47–49, 56, 57, 59, 75, 77, 92, 134, 136, 143, 144, 167, 168, 182, 186
Conservation of energy, 13, 14, 21, 23, 28, 167
Conservation of mass, 13, 15, 16, 20, 27, 100
Conservation of momentum, 13, 14, 17, 28, 167
Conservation vector, 13, 26, 68
Conservative form, 19, 76, 77
Consistency, 47, 59, 60
Contact surface, 106, 108
Contour, 71, 100, 120, 147, 168, 169, 171, 172, 187, 188
Control volume, 14–16
Convection, 23
Convergence, 57, 59, 60, 64, 87, 89, 90, 117, 119, 134, 136, 137, 143–145, 167, 168, 186, 187
Coordinates, 15–17, 20, 23, 25–27, 71, 73–77, 90, 115, 136, 137, 152
Coordinate transformation, 73
Courant-Friedrichs-Levy (CFL) number, 65